WE ARE THE SPEARS OF VICTORY

WE ARE THE SPEARS OF VICTORY

By: Kofi Piesie

© Kofi Piesie Research Team. © Same Tree Different Branch

Kofi Piesie/Mossi Warrior Clan

Copyright 2020 by Kofi Piesie Research Team

All right reserved. No Part of this book may be reproduced or transmitted in any form or by any means, electronic or mechanical, including photocopying, recordings, or by any information storage and retrieval systems without the written permission of the publisher.

Printed in the United States of America

WE ARE THE SPEARS OF VICTORY

Table of Contents

Foreward – Bro Lavelle 11

Dedicated Foreward – Robert McCardell...... 16

Preface – Chavis Tp-hsb McCray 29

Introduction – Ini Herit Shawn P 38

Survey One – The Knockout to Racism White Supremacy By: Sutekh Akande Iyanu Oluwa 44

Survey Two – Po Tolo By: Chavis Tp-hsb McCray 69

Survey Three - Syncretism: A Means and Method for Accurately Translating km.t, km.tyw, and km (Revisited) By: Ini Herit Shawn P.. 104

Survey Four – The Battle of Adwa By: Kofi Piesie 136

Survey Five - Product of the 44: Extraction of liberation and empowerment through Houston's Historical African American Enclave By: Chavis Tp-Hsb McCray........................ 157

Survey Six - The Attack on Intellectualism During the Age of Information By: Ini Herit Shawn P 177

Acknowledgment

Kofi Piesie

First, I would like to thank and extend honor and praises to my Ancestors who came before us, and second, I would like to thank those who have supported Kofi Piesie Research Team on our journey of trying to liberate the minds of our people.

Acknowledgment

Ini Herit Shawn P

I would like to thank everyone who has supported the Kofi Piesie Research Team and our Spears series. Hopefully we have helped you develop the necessary tools to reclaim your history, sense of self, and love for who you are. I would also like to thank my family, friends, and acquaintances who have helped motivate me to become a better student and writer.

Acknowledgment

Chavis Tp-hsb McCray

I'd like to again acknowledge my wife and kids who are always my motivation to be the best man I can be. I want to acknowledge my Granny Verda Hawkins who gets to see her grandson's name in a book again. I want to recognize my mom and dad for giving me this life to be able to write in these books and challenging me to be an intellectual. I want to acknowledge my master teacher and aunt Velicia Moore for inspiring me to put this work in for my people. I want to acknowledge my grandfather for his support. Long live my papa James "Duke" McCray and grandmother Sarah Mae McCray as well as my cousin Marlon "EMG Santana."

Gomez. Thank papa James Johnson for sharing his knowledge with me and Big Mama Ella Johnson for always loving me and just being the beautiful soul she is. Shout out to my boy BG Broderick Green who is free. Shout out to that boy Dre who is always very encouraging at work along with my supervisor Daniel. Shout

out to Neno Nelson Hughes and Tee, my Southside Ridgemont PUD Park big brothers from another mother. My Scarborough, Forest Brook, and Serra High alumni. Cousin Jackie, Shynedra, and Jas Perkins, sister-in-law Rae Rae, Glenda, Mardie, Mesha, Tony, CeCe, Robert, Deneshia, Jasmine, and special shot out to My cousin Delylnne Ellis and her husband along with my aunt Angie, Jean, and Leianna my beautiful auntie that is younger than me, my uncles Doug, Cooper, and Mike, Uncle Travis, Aunt Rena, Chaz, Jas, Chelsie, John, Sharell, Marlon and Lil John, Javion, Jordan, Aria, and that super awesome nephew Landon, Shane and Shante, for their support and. Free JD and RaRa. Long Live Valerie Hawkins, Shamecca Davis, Mama Karen McCray. I say all of your names to give you all immortality living forever in this text.

Acknowledgment

Sutekh Akande Iyanu Oluwa

As I am honored to still be a member of this intelligent group of Black men, I first want to thank the entire team. From T'Challa, to Setepenra, to Chavis, to Shawn and Kofi. I receive inspiration in different forms from each one of you. I would also like to give metaphorical flowers to our extremely loyal extended family. That goes from supporters who tune in every time we live stream to those who support Same Tree Different Branch Publishing and Mboka Brand Apparel. Each and every one of you guys are a part of the history that we are making and you are the reason that we do this. I also want to big up those that hold to pseudoscience and pseudohistory because they also inspire us in a different way. We, as a team, will continue to push forward.

"If you want to go fast, go alone. If you want to go far, go together."

Teams wins

Spears are flying, and pseudos are crying

WE ARE THE SPEARS OF VICTORY

FOREWORD

"Learn how to think ahead, so I fight with my pen."-Tupac Shakur (To Live and Die in L.A.)

By Bro Lavelle

"We can only understand our oppressors' psychology by understanding their history. They rob us of a knowledge of history and want us to think that history is irrelevant and unimportant so that we cannot see through their deadly games." (Wilson, Amos:18:1993)

After four hundred years of being stripped of our history, Black people have been in a loophole to where we as a people prefer to be anything but African. This disdain for Africa is real in 2022. Now I do recognize that not all Black people come from Africa, which would be a small amount, but the bulk of us come from Africa, and we abhor it. (Not all of us do.)

So, when my good brother Kofi inquired me to write the foreword for this significant edition, I was delighted and willing to uphold Africa's vibrant history as an African Warrior Scholar. For us to get triumph over our oppressors, we must study and appreciate our history as African descendants. "Most Negroes don't know history." (Malcolm X, 30:1971.) What Malcolm said is precise even until this very day.

The passion for Africa must be installed in us, and I honestly feel like this volume can motivate the reader to return to this beautiful, black landmass. This mental action, from my impression, would be called "Sankofa," which implies to look back and fetch. Glance back and recognize how great our predecessors were. We as a people keep this combination of prominence inside of us only if we can understand that and put it to use as part of the victory process.

In the 21st century, our commitment as African Warrior Scholars is to generate quality evidence so that the following generation can keep up this remarkable task that our African Scholars showed before us. This edition will incite the younger generation to perform other studying as the times that we are living in African history is being seized out of public schools and replaced with different subjects which serves more harm to our youth.

Brothers and sisters, dear readers, African liberation is not just taking on arms, but it contains picking up a pen and formulating words together and utilizing it to battle for the hearts and minds of our people. We are not just quarreling with our open enemy with their misinformation, but we also fight our people with incorrect data. As I keep asserting, this book will look to shatter the restraints off our peoples' minds with exact and solid methods.

Upon the completion of this book, my suggestion to the individual would be to pass it on to the following person so that they can be educated. Again, I want to thank Brother Kofi Piesie for letting me be part of this publication, and thanks to the readers who bought this as well. PEACE!!!

Brother Lavelle

SOURCES

TUPAC SHAKUR (TO LIVE AND DIE IN L.A., 1996)

MALCOLM X (THE END OF WHITE WORLD SUPREMACY: FOUR SPEECHES: IMAM BENJAMIN KARIM,1971)

DR. AMOS WILSON (THE FALSIFICATION OF AFRIKAN CONSCIOUSNESS EUROCENTRIC HISTORY, PSYCHIATRY AND THE POLITICS OF WHITE SUPREMACY,1993)

DEDICATED FOREWORD

Victory

Robert McCardell

I've followed the Mossi Warrior Clan for quite some time. I've appreciated the informative content that has been geared towards a reexamination of Africa in a different light from the past, where opinions were shaped by paternalistic, colonialist, and imperialist attitudes towards the so-called dark continent. Naturally, when given the opportunity to contribute to an upcoming volume, I was ecstatic about the possibility. Brother Kofi mentioned the title, "We Are the Spears of Victory," and I thought long and hard about the lessons that can be gleaned from the past that can be re-examined and offer potential insights.

The spear is a global symbol. In Africa, as in many places in the past, it evoked concepts of power and authority. As an instrument of war, the spear has played a pivotal role in gaining the tactical advantage in battles that have decided the wealth and domain of kingdoms. When considering the "Spears of Victory," there is one particular example from history that comes to mind that highlights the transformative power of the spear. It's the story of a military visionary who, with his spear, combined intent, discipline, skill, craftsmanship, and strategic intelligence to build one of the most renowned kingdoms in Southeast Africa in the 19th century under the leadership of kaSenzangakhona A.K.A "Shaka Zulu.", the Nguni tribe's people who eventually became known as the Zulu, represented a dreadful and powerful force in South Africa. The Zulu grew from a tribe of 1500 people in a small territory of 26 kilometers to over 250,000 people and an army of 40,000 warriors. Traditional warfare in South Africa was much more ceremonial than combat driven. There were battle dances similar to that of the Samoan Haka. Perhaps there would even be a showing of strength by way of stick jousting between members from rival factions. Sometimes the warriors would toss a lightweight, nearly 6 ft long (1.8 meters) throwing spear known as the Assesagi at opposing tribes. It was rare that the Assesagi would actually strike and kill a member of a rival faction. Due to the arc of the throw and the distance, soldiers could easily sidestep a thrown spear in most instances. If we had to label the Assesagi within the context of this idea of the "spears of victory," we could call the Assesagi the spear of peace, the spear of freedom. However, when European colonialism was fastly encroaching upon KwaZulu-Natal, it could be argued that Shaka Zulus's spear was the right spear for his time.

One of the key factors of the Zulu military success under Shaka was the reengineering of an existing technology, a long-tipped double-edged short spear that became known as the "ikwla". The "ikwla" is said to be named after the sound the spear makes as it exits the body. The "ikwla" was a work of art in every way to the extent that we can admire a weapon. From the skillfully crafted double-edged Iron tip, so sharp that it could also act as a sword, to the finely carved wooden handle that most times flared at the hilt, the ikwla was as beautiful as it was deadly in the hands of a skilled warrior.

Contrasting the traditional throwing spear (Assesagi) against Shaka's chosen spear (ikwla), it's symbolic of the shift in culture that Shaka ushered in. War was no longer limited to the realm of play. No longer would the spear be a weapon of chance to merely be tossed about as a projectile with little chance of reaching its target. The spear is now a weapon of intent to be used in close-quarters combat. Shaka combined his spear with the long shield known as the Isihlangu, which means to brush aside. In the hands of a skilled warrior, the Isihlangu could hook the shield of the opponent to disarm them while the IXkwa delivered the fatal blow.

Another pillar of Shaka's success was discipline. The Zulus trained in martial arts from a very young age. Shaka had a mentor system that allowed younger or less battle-tested soldiers to shadow seasoned veterans. The men were sworn to celibacy, believing that sex made them weaker in the heat of battle. Shaka also instilled a mindsight that reinforced courage and valor. He created a culture of winning. Men who were stabbed in the back or lost their spear were considered cowards because it appeared as if they either stood back or ran away from the battle. Men sporting war wounds in the front of their bodies were considered decorated soldiers. Shaka abolished sandals from his army; he felt that sandals slowed the men down on foot. Every Zulu warrior was barefooted, yet they were battle toughened to the point that they could traverse various terrain while unaided by any type of footwear.

Shaka was a military, and strategic genius, one of his most famous battle tactics was "the horns of the bull" or the crescent. The "horns of the bull formation" can be seen as a buffalo's body, emphasizing a central "chest" of seasoned warriors that engaged the enemy directly. The chest would act as a decoy, the enemy would naturally advance towards the chest, and in the heat of battle, the chest would withdraw from the enemy, luring them deeper into hostile territory. The left and right horns of the bull were made up of younger warriors who encircled the enemy while attacking them from the sides, preventing the enemy's ability to retreat. Back towards the body of the buffalo (center of the formation), but opposite of the buffalo's "chest" were the "loins," the men who protected the rear or flank. Shaka was known to personally assess the field of battle before each squirmish.

WE ARE THE SPEARS OF VICTORY

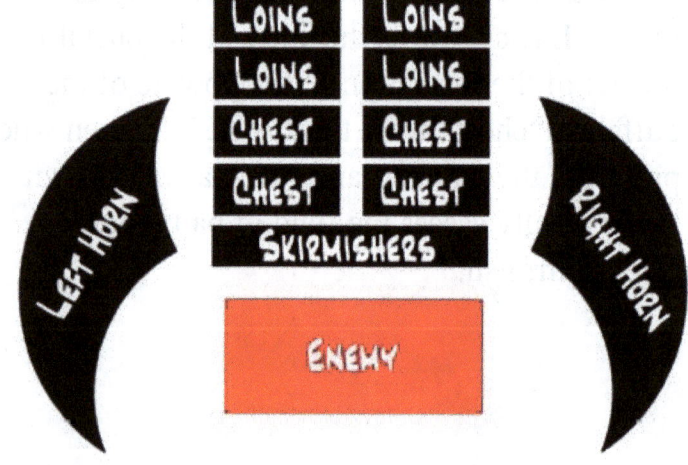

The skill involved in creating Shaka's spear is not to be overlooked or ignored. All Zulu spears were designed by craftsmen. When Shaka came along, Ironworking in South Africa was well established. High-quality steel was usually acquired through trade with Europeans, specifically the Portuguese. Spear makers were granted cows for high-quality spears, and it wasn't uncommon for a well-kept Ikwla to be passed down through the family.

When considering Shaka's spear of victory, I believe there are many concepts that we can build upon. Previous to the rise of Shaka, The Zulu practiced war as a game. The war games might've been useful to settle internal disputes or reduce unnecessary bloodshed. However, they were definitely insufficient to prepare for a formidable opponent such as the British. For a very long time the diaspora has been relegated to war games. We make menacing gestures, and we toss ideological weapons at the opposition; the rewards, more often than not, are empty or token victories. What would it look like in today's society for us to embrace the symbolism behind Shaka's spear? One of the most impactful things Shaka did in building a winning culture was to train the youth early to develop skills, discipline, physical and mental toughness to become effective warriors. Shaka's approach was structured, the youth shadowed the elders, and they learned by example. Emphasizing skill in strategy, crafts and trades are some things that we see in Shaka's victories. Blacksmithing was typically passed down in the family. Blacksmiths were revered as men who literally wielded forces of nature. A clear takeaway here is rebuilding the culture that was lost through oppression. A culture that encourages learning skills through knowledge and apprenticeship that can lead to self-sufficiency. As a leader, Shaka was very effective at building coalitions. The buy-in from the Zulu clans was not facilitated simply by sheer force alone. Shaka empowered Chiefdoms, decorated soldiers, and weapons makers via a meritocracy that was incentivized through land and cattle. How could we incentivize a coalition today? It's imperative to bring the information to the people, but a coalition is held together by some tangible mutual benefit, be it currency or safety, but most certainly continued group progress.

Ultimately, the spears of victory can be summarized as a concerted effort to build and support a culture where every aspect is geared towards the common goal of winning. Every member of the culture should be a stakeholder and should have some notion of the necessary sacrifice that must be made to achieve the ultimate desire. Furthermore, the culture must understand that sacrifices are often made to lay the groundwork for future progress, so there may not be an immediate return on investment. The celebratory aspects of the culture should be positive reinforcement for the habits that have led to success. Finally, each member of the culture should respect accountability and be prepared to hold others accountable.

SOURCES

https://ageofrevolutions.com/2019/04/15/the-zulu-iklwa-evidence-of-an-african-military-revolution-in-the-nineteenth-century/

https://www.metmuseum.org/toah/hd/iron/hd_iron.htm

Zulu IKLWA

https://ageofrevolutions.com/2021/01/27/born-out-of-shakas-spear-the-zulu-iklwa-and-perceptions-of-military-revolution-in-the-nineteenth-century/

https://ageofrevolutions.com/2019/04/15/the-zulu-iklwa-evidence-of-an-african-military-revolution-in-the-nineteenth-century/

http://web.prm.ox.ac.uk/weapons/index.php/tour-by-region/oceania/africa/arms-and-armour-africa-39/

https://www.tota.world/article/3108/

https://www.sahistory.org.za/people/shaka-zulu 10/23/2022

https://ageofrevolutions.com/2021/01/27/born-out-of-shakas-spear-the-zulu-iklwa-and-perceptions-of-military-revolution-in-the-nineteenth-century/

https://www.britannica.com/biography/Shaka-Zulu-chief

https://curioushistorian.com/the-rise-of-shaka-and-the-zulu-empire

chrome-extension://efaidnbmnnnibpcajpcglclefindmkaj/https://www.anglozuluwar.com/images/Journal%2043/Identifying%20and%20dating%20Zulu%20spears.pdf

https://moguldom.com/57305/16-things-made-shaka-zulu-a-military-genius/6/

Deflem, Mathieu. "Warfare, Political Leadership, and State Formation: The Case of the Zulu Kingdom, 1808-1879." Ethnology, vol. 38, no. 4, 1999, pp. 371–91. JSTOR, https://doi.org/10.2307/3773913. Accessed 30 Oct. 2022.

https://www.researchgate.net/publication/301375273_Applicability_of_Shaka_Zulu's_Leadership_and_Strategies_to_Business

https://adf-magazine.com/2021/12/shaka-zulu-and-his-deadly-spear/

Preface

Spears of Victory can be interpreted literally and figuratively as a weapon utilized to not only combat but conquer on the battlefield. Whether that battlefield be a literal, specific location or a figurative, intellectual, or social environment, one must be well armed. KPRT aim is to ensure that our people are properly prepared and well-equipped with historical context, sound methodology, bulletproof logic, and basic competency in said subject matter.

Pic of Nwanyeruwa retrieved from

https://www.blackpast.org/wp-content/uploads/Nwanyeruwa-public-domain.jpeg

Drawing from our rich African history and culture, we are setting the standard to compile our great ancestor's wisdom, knowledge, and history in book form to give them immortality. We strive to educate the youth on rebellious warriors of our past like our great ancestor Nwanyeruwa who was an Igbo woman from a village called Oloko in Southeastern Nigeria that organized a woman's revolt against taxation imposed by the British colonial administration and had a significant leadership role in the Aba Woman's war. (Ezeh, 2022)

Blackpast. Org informs us, " Upon hearing that British colonial authorities who governed Nigeria at that time were attempting to implement new census and taxation measures on women in her region, Nwanyereuwa took action that changed the course of Nigerian history. She utilized the market networks common in traditional Igbo society to organize a women's movement against local Igbo warrant chiefs, particularly one Chief Okugo. Warrant chiefs had long been the main collaborators with British colonial authorities and routinely enforced their policies" (Ezeh, 2022)

Today we still deal with the oppressive introjected sentiment and behavior of our own people, enforcement of not only foreign policies but foreign values, and oppressive reasoning. Kofi Piesie Research Team aims to change the course of African American history, challenging false faulty narratives and misinformation while at the same time edifying readers through documentation of little-known history, offering context where it has been oversimplified and glossed over. Each volume we put together represents our attempt at claiming victory regarding our mission. Every volume is a demonstration of our research team accomplishing our goal of keeping our "As I learn, we all learn" motto a reality.

Just as Nwanyereuwa was pivotal in standing up to colonialism and those trying to oppress her people, we make it our priority to stand up to blatant racist outdated as well as current research. We are Afrocentric scholastic warriors and intellectual gunslingers who hold African autonomy in high regard, refusing to accept misrepresented and unverified racist and oppressive opinionated data objectively denying feel-good and exaggerated blackologist pseudo-scholarship.

I take pride in being a part of the brotherhood that is Kofi Piesie Research Team. A group of brothers that decided to make right knowledge and methodological approaches to research their prerogative. The spirit of the American Negro Academy lives through us as we continue the work they began. We are the New American Negro Academy, and we welcome you to yet another priceless publication filled with well-thought-out surveys intended to advance African American scholarship and correct it. None of this work is ego-driven. The inspiration is motivated by the fact that the work needs to be done, and we do not mind lacing up our scholastic boots.

Pic of Crummel retrieved from
https://en.m.wikipedia.org/wiki/File:Alexander_Crummel_(cropped).png

Kofi is like a modern-day Alexander Crummel, and our research team holds the same "promotion of literature, science, and art; the culture of intellectual taste; the fostering of higher education; the publication of scholarly work; the defense of the Negro against vicious assaults." (Britannica, 2014) objective as the original group.

We intend to continue to publish works and clog up the social media pipeline and abroad with honest research that correlates with our objective. Some may view our works as non-groundbreaking and feel we are study buddies that are not making any real contribution to our community, but the objective truth is contrary to that view. This information is groundbreaking to the individual unaware of the information in our books. We are making a contribution to the Afrocentric collective consciousness by default with the mere act of taking the initiative to document information and publish it for edification and if necessary, correction. We are not above reproach, nor do we make claims to identify as experts. We regularly defer to experts vs. solely appealing to them as the ultimate authority. The brothers of KPRT and our affiliates do our homework and share it with the class. We invite critique and embrace correction where it belongs. Feel free to at any time give us feedback on what you read in our books. We look forward to it.

Sources

The Heroine Collective (n.d) "Nwanyeruwa ", Retrieved from

http://www.theheroinecollective.com/nwanyeruwa-the-womens-revolt-against-british-colonialism/

"The Colonial and Pre-Colonial Eras in Nigeria"

https://www.historians.org/teaching-and-learning/teaching-resources-for-historians/teaching-and-learning-in-the-digital-age/through-the-lens-of-history-biafra-nigeria-the-west-and-the-world/the-colonial-and-pre-colonial-eras-in-nigeria/the-woymens-market-rrttebellion-of-1929

NWANYEREUWA
https://www.blackpast.org/global-african-history/nwanyereuwa/

Judith Van Allen (2014) "Sitting on a Man": Colonialism and the Lost Political Institutions of Igbo Women"

https://www.tandfonline.com/doi/abs/10.1080/00083968.1972.10803664

"Wicked" Women and the Reconfiguration of Gender in Africa (review)

Felix K. Ekechi

Africa Today

Indiana University Press

Volume 50, Number 1, Spring 2003

pp. 131-133

10.1353/at.2003.0056

https://muse.jhu.edu/article/47201

Britannica, The Editors of Encyclopedia. "American Negro Academy". Encyclopedia Britannica, 28 Aug. 2014, https://www.britannica.com/topic/American-Negro-Academy. Accessed 11 October 2022.

Introduction

Before the ceremonial war chants and ancestral reverences, the warriors gather to reassure themselves that whatever enemy presents itself shall fall for the betterment of its people. The elders and wives prepare the sacrificial meals. At the same time, the warriors paint their faces in war paint created by the diviners who foretell stories of old regarding great warriors of the past who drank from sacrificial cups of the ancestors. While blacksmiths arm the warriors with Shields and Spears, each warrior is handed a special war mask made with the blessings of the spiritual world. The war drums are playing as the warriors reveal themselves to a crowd blessing them with war chants and blowing the gods' smoke all over them. As the tribe elders approach the ceremony center, the crowd begins to silence. Words are spoken in the direction of the warriors, and calm sets over the ceremony.

The sacred spear is handed to the warriors and passed around as they pay respects to their ancestors. A roar erupts from the warriors, and the ritualistic fire caste reflects the warrior's readiness. Each warrior takes an oath as he acknowledges that any outcome, good or indifferent, is meant to be. They began to pound their weapons against their shields in solidarity, causing a tremendous echo throughout the camp. A new war chant is uttered as the warriors start to enter the forest to fight and the people bow in respect for the mission ahead is a long one. Silence takes over the crowd, and the elders summon the war deities to give them strength to be successful.

The enemy educated in pseudoscience, pseudo-history, and pseudo-linguistics makes themselves easy to spot. A signal is sent from the first warrior who emulates the Egyptian nTr, who reigns during the 18th Dynastic period. His success during the Battle of Kadesh instills trust within the group. His familiarity with hunting, fishing, gardening, unique mining currency, and weaponry provides the first significant blow to the enemy. As the warrior's advance, a loud echo from a war drum uncovers a potential trap. Still, a mighty Djembe drummer familiar with martial art and boxing style strikes the enemy with fierce blows that lay the enemy down quickly. The war drum again informs the other warriors it's okay to advance. As they proceed, another warrior notices a diviner calling upon the spirits of his ancestors to slow the attack.

Still, the intelligence of a mighty warrior with a sense of distinguishing spells and roots counters the diviner's attempt. He uses code-switching and proper methodology to clear the warriors' way as they advance their attack forward while the enemy mounts its following plot against them. He gives each warrior an herbal and purple liquid substance with instructions on when to drink it. This drink is to refute any potential threat of future spells from zealots.

The enemy on its heel, the leader of the tremendously skilled warriors, acknowledges that at some point, a political strategy will have to be deployed to end the war. Still, he must stay on the attack because any sign of weakness could lead to the destruction of his team and although the enemy may seem weak in some areas, its tactics are far deceiving. He calls on the elders and ancestors to keep him sharp as progress is made, and the enemy continues to become weaker and weaker as they advance.

As the enemy slowly retreats, they expose foreign pathogens in the atmosphere to all the warriors. Unfortunately, everyone had been prepared for this move foreseen by its most controversial warrior, who informed his team before they headed into the battlefield to ensure each of them got inoculated against these pathogens. Identifying the move immediately, he sees it as a sign of weakness, calling them cowards as he continues to throw spears. The enemy attempts to slow the warriors' progress but fails because their methodology is flawed, and due to familiarity with West Africa, Moorish history, and Quidah not being Judah, he made sure the team had enough instinct to recognize how the enemy would disguise themselves and hide in specific places to try and sneak attack them.

The trap that had been set wasn't adequately assembled, and now their leader stands surrounded by five heavily armed and protected warriors. But that wouldn't stop the pseudo attempts of the enemy to assert claims they cannot prove or haven't thoroughly investigated. As the warriors look amongst each other, the eldest of them approaches the leader to negotiate the terms of a potential treaty. This treaty would highlight the greatness of African traditions, customs, Kimoyo, art, history, science, culture, languages, and proper methodology. It would promote a cohesive inclusion to dispel myths and pseudoisms. It requires a blood oath of the enemy, and refusal of this treaty would mean death. Victory would be established by way of the Spear or Pen...... As order is re-established and harmony is restored.

SURVEY ONE
The Knockout to Racism White Supremacy

The Knockout to Racism White Supremacy
Sutekh Akande Iyanu-Oluwa

Racism has existed in all forms and aspects of life. Sports has been, but one of the vehicles used to push the agenda of White supremacy. Preventing African Americans from the highest prized accolades in sports, subjecting them to physical and verbal abuse, and barring them from participating in national leagues and organizations with Whites completely. We need not look no further than such instances and people like Jackie Robinson's plight of joining Major League Baseball. Bill Russell's home being broken into by angry racist fans to vandalize his home, spray paint racist terms on his living room walls and defecate in his bed. Lee Elder, the first African American to participate in the Masters of the Augusta National Golf Club, had to rent two houses during the tournament to safeguard himself from racist attacks. Probably the most prominent person in sports with the loudest mouth is Muhammad Ali. Ali is a man that is loved and cherished in contemporary times. However, during his reign, he was despised by White America for his socio-political stances outside of the ring. But Ali had a huge inspiration that came before him, in the form of legendary Jack Johnson, that motivated him to stand strong in the face of racial oppression and injustice, as Ali publicly praised Johnson at any opportunity he received.

"Jack" Johnson was born John Arthur Johnson to emancipated slaves in Galveston, Texas, on March 31, 1878 (Orbach, 2020) (Gustkey, 1990) (Flatter). Jack was one of nine of his janitor father's children (Gustkey, 1990). This was, without saying, a time of hardship and poverty for most African Americans that remained in those southern states after the emancipation from chattel slavery. Though a young Johnson made the best of a dire era of disparity. The hardship of residing in such a tough neighborhood as a youth benefitted him in learning how to defend himself (Gustkey, 1990). This was a classic case of turning lemons into lemonade, as a young Jack Johnson used the skills he learned from fighting in an uncivil area and turned them into legendary pugilism.

Jack proved to be a child of handiness as he spent much of his early years working on sculleries and boats in Galveston (Flatter). He would not remain in Texas for too long as he left Galveston at 12 years of age. He caught a train and journeyed across the southern states. Not too soon after that, Jack began to attend training camps for notable Black boxers in his mid-teens. He eventually became a sparring partner (Gustkey, 1990). Realizing that being a fighter was something that he seemed to excel at, this would only be the birth of the "Galveston Giant" and the start of his transcendency as a boxing icon and innovator.

When he was 17 years old, Jack decided to return to his hometown. He happened to catch the traveling circus one day when they came to Galveston with their Carnival fighter, Bob Tomlinson. Bob would bet $5 to anyone that could hold their own and last 4 rounds with him (Gustkey, 1990). Unfortunately, Tomlinson underestimated Jack and got an unexpected surprise. This shock was in the form of a clean knockout in the fourth and final round (Gustkey, 1990). This was the moment that made Jack start to take pugilism more seriously. It was the turning point that sparked the idea that, just maybe, boxing was his calling, despite the racial disparities of the time.

The big fellow, in due course, sprouted to 6 feet 1 inches, hence the nickname "Galveston Giant" (Flatter, ESPN). Migrating to Chicago to further pursue his goal of being a professional boxer and rising up the rankings to become a champion, he initially fought for small purses. A "purse" in the context of boxing is a set amount of money, agreed on by both fighters before a fight, to be paid out at the end of the fight. Jack would start to second guess his goals, not because of a lack of confidence in himself, but in the racist White supremacy in America in allowing an African American boxer to shine and make it big time (Gustkey, 1990). Jack was a pioneer, so he had no boxing influence to idolize or look up to that had broken the "color lines" in boxing. Little did he know that he would not only be the one in boxing to break the "color lines," but he would be the one to knock its teeth out.

Around 1901, Jack left Chicago, Illinois, and moved to California to provide more fight opportunities for himself (Flatter, ESPN). He fought in cities such as Los Angeles,

Bakersfield, San Francisco, and Oakland in a span of three years (Gustkey, 1990). It seemed as if Johnson's hard work and dedication were beginning to pay dividends while in California. He would get a title shot at the Colored Heavyweight Championship of the world on February 3, 1903, in Los Angeles. He was victorious in a 20-round decision versus his opponent in Denver, Ed Martin (Flatter, ESPN). This would prove a major steppingstone for Jack, but his aim was much higher. He dreamed of not only being the Colored Heavyweight Champ but crossing that racial barrier and becoming the Heavyweight Champ of all world races.

Photo of James Jeffries

The following year, in 1905, the current and reigning undefeated Heavyweight Champ of the World, James J. Jeffries, retired because there were no more worthy White opponents to fight, and Jeffries refused to give a title shot to a Black man (Orbach, 2020, p.280). Back in March 1898, James defeated Peter "Black Prince" Jackson in knockout fashion. Peter Jackson was known, at that time, for being the African American boxer that White boxers, such as John L. Sullivan, would "duck" because they rejected the idea of fighting a Black fighter. Sullivan was, without a doubt, called out on attempting to rationalize claiming to be the best but not fighting the best, as Peter Jackson was the most acclaimed fighter of the 1880's (Orbach, 2020, p. 280). As one can see, well before the time of said boxers, a color line had been drawn in the sand. This, essentially, made the sport of boxing somewhat segregated.

Jeffries was young, hungry, and in his prime. As a result, he had no issue getting into the squared circle with Black fighters, as he was on his way to the top to become the number one contender for a heavyweight title shot; However, the Black boxers that he fought, like Peter Jackson, were old and past the prime of their careers (Orbach, 2020, p. 280). He gained the World Heavyweight Championship at the young age of 24, on June 9, 1899. Jeffries then refused to give any Black fighters an opportunity after gaining heavyweight title success, including the fast-rising star Jack Johnson (Orbach, 2020, p. 280). Jack was thirsty for a shot at the title and even knocked out James Jeffries' brother, Jack Jeffries, while making his way up the rankings. He began publicly calling for this title fight with undefeated champ James Jeffries in 1904. "I want Mr. Jeffries next. I think I am entitled to a fight with him. I am faster than ever and bigger and stronger. I guess everybody knows it", Jack told the press (Orbach, 2020, p. 283).

Due to the difficult time he had finding White competition of merit and knocking out mediocre opponents in the initial rounds, in August 1903, Jeffries declared his retirement from boxing when there were no more worthy White men to fight (Orbach, 2020, p. 281). This was an intentional ploy to keep heavyweight division boxers of African descent stagnant. This is shown from the act of some White boxers, like Jeffries, getting into the ring with their Black counterparts when convenient to them as they maneuver their way to the top while not privileging Black boxers with a shot at the most prestigious title in boxing once they are on the throne.

Boxing commentators thoroughly attacked white Champions for circumventing the Blacks to retain their titles. Jeffries was a strong supporter of this, saying that he would not compete against any Black boxers unless it was demanded by the wider public (Orbach, 2020, p. 282). This was a call to White America as a whole because Black fight fans had every intention on the great Jack Johnson becoming the lineal heavyweight champ. This critique of White champs intensified when, in April 1904, James Jefferies officially announced that he had one year before retirement and still refused to fight Johnson. But Jeffries went back on his word, as he continued to decline a match to Jack Johnson, albeit the public demanded it. The color line was firmly but carefully drawn (Orbach, 2020, p.284). interracial fights weren't just fights. They held racial connotations as a win for a White boxer against a Black boxer would metaphorically strengthen the idea of White supremacy, and a win for the Black fighter against the White fighter would metaphorically support Black supremacy in the minds of White America (Housman, 2021).

James Jeffries vacated his title to the winner of a match held in Reno on July 3, 1905, against Marvin Hart vs. Jack Root, after retiring prior in that same year. A 12th-round K.O. by Marvin Hart sealed the deal, but in February of the following year, he lost the belt to Tommy Burns. Burns would prove to be the final obstacle in Jack Johnson's road to Champ (Orbach, 2020, p. 285).

Jack would finally get his opportunity via Tommy Burns, three years after the retirement of James Jeffries on December 26, 1908. This event would take place in Sydney, Australia. The 14 round bout concluded with Tommy Burns being knocked out by The Galveston Giant, Jack Johnson (Housman, 2021). He had finally done it. Jack had become the heavyweight Champ of the world. The first and only Negro to do so at the time.

The aftermath of this title switch led White Americans on a search for a "Great White Hope," forthwith, as an answer to Jack Johnson. To them, having a Black man hold and defend the most coveted title in boxing was impermissible in the era of segregation, Jim Crow, and White authoritarianism (Housman, 2021).

Photo of Tommy Burns vs. Jack Johnson, December 26, 1908, in Sydney, Australia

Jack was on top of the world. He was known as a "bad nigger" because he refused to comply with the racial and social standards of the time. He also dated and married White women, which caused rage in White America at a time when interracial mingling was heavily frowned upon (Orbach, 2020, p. 273). Jack defied all odds placed on him by White supremacists. During an epoch in American history of Black fighters being pseudo-scientifically coined as unintelligent, solely based on the color of their skin. Jack had a high ring IQ and was a master defensive strategist. He seemed elusive at times, the way he fatigued his opposition by slipping jabs and parrying attacks to wear them down for the knockout in the later rounds. Most Black boxers of the time didn't fight White boxers with the same intensity and effort. Jack not only was the exception to this rule, but he humiliated, taunted, and laughed at them before, during, and after the fights (Gustkey, 1990).

Johnson's next battle of note was on October 16, 1909, in Colma, California. Almost a year after the fight that made him Heavyweight champ, Jack stepped into the ring with Stanley Ketchel, middleweight champion. Jack clearly had the size advantage, being over 6ft tall and weighing in at 209 lbs., over Ketchel's 5'9", 160 lbs. (Bearden, 2006). This was a mismatch. Jack took advantage of this size difference early in the fight. Stanley Ketchel, "The Michigan Assassin," must've had his confidence on one thousand to even consider taking this fight. As a fan of boxing, I understand how detrimental this discrepancy is. By the time they entered the 12th round, Ketchel had been battered, bloodied, and abused. But to Johnson's surprise, after being knocked down in the 2nd round, Ketchel hit Jack with a punch that dropped him to the canvas. After a brief recovery after the count, in an angry rage, Johnson severely knocked out Ketchel with a punch so hard that it broke multiple of Stanley's teeth. After this K.O., Jack had to remove the challenger's teeth that were lodged in the Champ's right glove (Bearden, 2006).

This brutal knockout would serve as the final straw that made the retired undefeated Champion, James Jeffries, want to defend the White race against Jack Johnson. But by then, he hadn't fought in years and was out of shape. He had let himself go, living in retirement on his alfalfa farm. By mid-1905, he was recorded as weighing 314 pounds (Gustkey, 1990). Months before Johnson vs. Jeffries' fight of the century, Jeffries told the media, "I feel obligated to the sporting public to at least make an effort to reclaim the heavyweight championship for the whole race. I should step in the ring again and demonstrate that a White man is the king of them all" (Orbach, 2020, p. 272).

Jack was immediately the big underdog as they headed into the fight. Articles, such as an article in Current Literature Magazine, "The Psychology of the Prize Fight:," gave pseudo-scientific backing for predictions of Jeffries's win. According to this article, Jeffries was blessed with Whiteness, which equals intellectual superiority, while in turn, Johnson's Blackness would be the cause for emotional streaks (Orbach, 2020, p. 273). Other publications, such as Harper's Weekly, discussed the financial aspects of the fight and how the most important part of the economics would be the motion picture film that would be shown worldwide. They also speculated that there was a possibility of racism harming the revenue to be made if some states and municipalities refuse to screen the fight (Orbach, 2020, p. 274).

The purse offered to Jeffries made the decision to fight even greater. The promoter offered a guaranteed $120,000 and just short of one million additional dollars if Jeffries was to win. When they weighed- in, Jack Johnson was 212 pounds and Jeffries at 227 pounds. Jack whooped on power-puncher Jeffries. His time out of the ring had him rusty. Champ Jack would make him catch punches as he talked to Jeffries and even to people at ringside. "Hey Farm, ever see a champ eat leather? You just watch Jeff. He loves to swallow mouthfuls all day long, don't ya, champ?" Jack yelled mid-round to Jeffries's trainer, Farmer Brown (Gustkey, 1990).

By the 15th round, James was wounded and bleeding from cuts over his eye and in his mouth, while his opposite eye was almost swollen shut. Jeffries got caught and knocked down with a combination. As he rose up to beat the referee's count, Jack went on the attack to finish him off and knocked Jeffries halfway through the ropes and onto his back. Again, Jeffries beat the count on shaky legs and attempted to dodge to the opposite side of the ring to avoid Johnson. This is where he was knocked down for the final time. Before then, the ref counted him out, and his team stormed into the ring (Gustkey, 1990). This would technically be a disqualification for Jeffries, although it is recorded as a knockout. "No, Jack, don't hit him anymore!" was the outcry from James' manager, Sam Berger, as he hurried into the ring (Gustkey, 1990).

Jack Johnson knocks James Jeffries down and halfway through the ropes.

Once news of the results of the fight spread, via new media, across the country, White America was enraged that their ideology of supremacy was challenged and beaten. Their last hope had been knocked out. Less than 24 hours after the fight, the infuriation of the outcome caused what was termed, racial "riots" in many major U.S. cities. Theresa Runstedtler, American University Associate Professor of History and author of "Jack Johnson, Rebel Sojourner: Boxing in the Shadow of the Global Color Line," says, "And they called them race riots, but essentially it was white mob violence against African Americans. It became this kind of attempt to put African Americans back in their place" (Housman, 2021). White America couldn't stand that their king had been demolished by someone of a race that was seen as less than human. Essentially, a Negro was the best fighter in the world.

A little more than an hour after the big fight results were announced on bulletin boards, 11 riots were reported in New York. One Black man was bludgeoned to death, and more than 100 were beaten up while Whites endured knife and bullet wounds. In Tenderloin, Manhattan, a Black man was captured by angry White men and hung up on a light post. He was nearly dead when police cut him down. On 135th/8th Avenue, a mob of angry White men stormed up to a car and pulled out a Black man into the street, and proceeded to kick and beat him before police rescued him. A Black woman was the victim of an attempted lynching while buying a newspaper. She held off the crowd wielding her stiletto shoe until the police arrived. The Negro section of San Juan Hill in Manhattan burned a Negro tenement house and threw big rocks at the windows. In an attempt to keep the occupants from escaping, they blocked the exits. Luckily the mob was dispersed by the fire department. It took the entire police department of Pueblo, Colorado, to end the Black vs. White conflict. After lynching 2 Blacks, Blacks and Whites were on the edge of a race war strengthened by the Johnson victory in Charleston, Missouri. Three Negroes were killed by race riots in Uvaldia, GA, near Augusta. In Atlanta, a Negro was jumped by 6 White men for yelling "Hurrah for Johnson!" on a busy downtown street but luckily, police rescued him. Two men were shot and killed in Washington. Two hospitals were crowded with injured and city jails were full of

236 rioters. Three riots were reported in Pittsburg. In Wilmington, Delaware, a mob of angry Blacks attacked a white man over an argument about the Johnson/Jeffries fight. A White mob then chased the Black mob and rushed the house that Blacks attempted to hide in. The 20,000 Black population of Columbus, Ohio, commemorated Jack's victory. Around 400 Blacks paraded down the city streets with a band. There was fighting during the marching. Few were hurt, but police intervened to stop the violence and allow the Blacks to continue celebrating (1970). There were 19 reported dead, and hundreds of others were injured (Housman, 2021).

In the immediate days after the "Fight of the Century," the New York Times published a leading article directed toward the Black population that celebrated Jack Johnson's victory. "Do not point your nose too high. Do not swell your chest too much. Do not boast too loudly. Do not be puffed up. Let your ambitions not be inordinate or take a wrong direction. Remember, you have done nothing at all. You are just the same member of society today you were last week. . . . You are on no higher plane, deserve no new consideration, and will get none" (Gustkey, 1990). The unprofessionalism was rampant. The anguish that White supremacists must have felt after they have placed all their hopes on Jeffries as their White savior, only to have him humiliated in the ring by a Negro.

Knockout and all, the biggest financial losers in the entire situation were those that purchased the rights to the fight film from promoter Tex Rickard and the fighters, as they planned on a Jeffries win which could be used as a $1 charge to view the film all over the world (Gustkey, 1990). To pacify any further violence, states and municipalities immediately started to ban viewings of the fight footage. This would start a precedence that would end with the Federal government banning all screenings of title fight footage in 1912. "White authorities were worried about the symbolic implications of the Jack Johnson victory being replayed …... They worried that any demonstration of Black victory, and any demonstration of white weakness or defeat, would undercut the narratives of white supremacy. Not just in the United States, but in colonies like South Africa, India, and the Philippines", says Theresa Runstedtler (Housman, 2021).

Two days after the fight, Jeffries was honest with himself and his White fans, "I could not have beaten Johnson on my best day," he told a reporter while on a train from Oakland to Los Angeles. Johnson returned to Chicago, where he would go broke in three years from his notorious spending habits. Hated ore by White America after marrying 2 more white women, Johnson's first wife committed suicide in 1912. The Federal government then began to find a way to target Johnson by trumping up charges against him on the Mann Act, which made it illegal to move women across state lines for "immoral purposes. This led him to be convicted and sentenced to one year in federal prison (Housman, 2021). A demonstration of the power of the American judicial system was flexed to send a message: You will lose one way or the other, and you will always be a nigger.

Jack escaped the U.S. disguised as a member of a Canadian baseball team. He moved from Europe to South America, Tiajuana, Mexico for 8 years before turning himself in at the US-Mexico border in San Diego in 1920. He served his year in captivity at Leavenworth penitentiary and spent the remainder of his life making public star appearances (Housman, 2021). On June 10, 1946, while driving from Texas to New York for the Joe Louis/Billy Conn title fight, Jack lost control of the car on a country highway close to Raleigh, North Carolina, and crashed into a power pole. He died shortly after the collision. At 68, he was buried in Chicago next to his first wife (Gustkey, 1990).

Jack Johnson revolutionized boxing by kicking down the door of white heavyweight champs that refused to fight Blacks to offer them a title shot. He became the first Heavyweight Champion of African descent and delivered a knockout blow to racism that can still be felt in sports today. He was a racist white man's nightmare. Big and brawn, intelligent, and flashy, he dated some of the most desired white women of the time, such as moulin rouge star Mistinguette and sex symbols Lupe Velez & Mae West. He owned a nightclub in Chicago and a jazz band. He would act on stage and is reported to walk around sipping on champagne while walking his pet leopard on a leash and showcasing his gold-toothed smile, which matched his gold-handled walking stick (Flatter). The flashy razzle dazzle of Johnson, and his embrace of the villain role (as Blacks are used to being portrayed as in modern sports era), and his defensive prowess brings to mind a Floyd Mayweather. He would retire with a record of 79 wins (46 by knockout), 8 losses, 12 draws, and 14 no-decisions (Flatter). True boxing fans and historians will forever recognize him as a man that opened the doors for all Black Heavyweight champions, including the great Muhammad Ali. Forever will he be labeled as the Black man that gave the knockout punch to White supremacy.

SURVEY TWO
Po Tolo

Polo Tolo

Chavis Tp-Hsb McCray

In this survey, I will document my social media experience on a controversial topic of the Dogon and cultural ethno-astronomy. My intention with documenting this presentation in literary form is to provide additional context to the position I took in a 3-hour long presentation that can be found on KOFI PIESIE TV YOUTUBE CHANNEL. Objectively I received mixed reviews on the presentation, and I was comfortable with in all honesty. I never had an issue agreeing to disagree.

Those who praised the presentation appreciated the in-depth information and sound methodological approach. They appreciated the sourced material that I left for further reading and edification as well as complimented our new visual presentation, eye candy with all the interactive video slides with stars shooting across the background, moving away from the dull PowerPoint slides changing up in Canva. This app lets you create book covers, videos, presentations, banners, logos, etc.

There was social media talk that someone did the presentation for me, which simply wasn't true. I had actually did the whole presentation on PowerPoint and still, have it saved to this day. My good brother Shawn of the MOSSI, SESHEW, and KPRT actually converted that presentation in canvas, trying to encourage me to make the switch to offer a more personalized and entertaining dynamic to what was characterized as boring readership. People don't know there were 100+ slides in that presentation. Shawn had to split it in 2 because it was so much information. The brother did his thing and converted me into a Canvas using somebody. That man is an editing machine and leading the charge in finding innovative ways to edify his people. If you don't know where to find these displays of brilliance, explore and subscribe to the Science with Shawn YouTube channel, where scientific literacy and competence is a priority and standard. Nobody is giving the people the most up-to-date happenings in the scientific community and abroad like him and the virology professor I introduced boys to by the name Vincent Racainello.

Getting back to the feedback, there were those who felt I struggled to read the presentation, and the truth is I honestly struggled to see the presentation because I hadn't received my new eyeglasses, so I was using my natural vision, which happens to be extremely poor 24/7 minus glasses. Shawn critiqued my laziness in my screenshot excerpts from my phone. He challenged me to either type it out or summarize it, which is advice I still use to this day. Those screenshots were tiny and difficult to see with my horrible vision, as well as difficult for the audience to make out. I didn't trip because I made sure to provide all sources in the description of the video. One of the more frustrating views was that basically nothing was demonstrated or proved, so this particular survey will aim to show why I agree to disagree with that notion in a clear and concise contextualized manner.

The most important part of that presentation that opposing views omitted was the research question: What is a star system?

Did the Dogon have knowledge of a star system? Did Griaule influence their understanding of a star system? What can be extracted from responses to Van Beek? I have adopted this method of making sure I lay out my research questions off the muscle from Shawn as well as it is a sound way to outline how you are going set out to answer these important questions and decide what evidence you are going to utilize to substantiate your proof and support your position.

What is a star system?

To attack this question, I utilize this quote from a learning website which stated "Although constellations have stars that usually only appear to be close together, stars may be found in the same portion of space. Stars that are grouped closely together are called star systems. Larger groups of hundreds of stars are called star clusters. The image is a famous star cluster classed M45, also known as Pleiades, which can be seen with the naked autumn sky.

Although the star humans know best is a single star, many stars—in fact, more than half of the bright stars in our galaxy—are star systems. A system of two stars orbiting each other is a binary star. A system with more than two stars orbiting each other is a multiple-star system. The stars in a binary or multiple-star system are

often so close together that they appear as one, and only through a telescope can the pair be distinguished."
(https://courses.lumenlearning.com/geophysical/chapter/star-systems/)

This description was important because it helped define star systems and clusters and explain binary and multiple Star systems. This criteria of what these things are helping answer the next research question involving whether the ethnic group identified as the Dogon had any knowledge of these concepts of their own.

Did the Dogon have knowledge of a star system?

To begin to dive into this inquiry, this quote, "Since ancient times, people all over the world have stared up at the night sky and struggled to make sense of it. This curiosity gave birth to one of science's oldest disciplines: astronomy – the study of celestial objects and phenomena. But astronomy is arguably more than the science of the stars.

It is intimately connected to our ideas of ourselves, our purpose, and place in the universe."

Urama, Johnson. "Celestial Stories." Indigenous and Cultural Astronomy, The ACU, 26 Apr. 2021, https://www.acu.ac.uk/the-acu-review/celestial-stories/.

By Johnson Ozoemenam Urama, University of Nigeria, Nsukka clarified that stargazing is an ancient and universal activity globally which would definitely and irrefutably qualify Dogon people by default, seeing as how they reside on this same globe as all the other stargazers. The quote contextualized how astronomy surpassed identifying or labeling stars and displayed connection to how we choose to do so. This sets up the alley-oop for how this involved the Dogon and their autonomy in their cosmological and astronomical indigenous understanding. Adding support to this notion, "Africa's astronomical systems are as old as its people. For thousands of years, observations of the cosmos have guided communities in practical and spiritual ways, from agriculture and navigation to religion and ritual. Indigenous communities were the keepers of a rich body of astronomical observations and knowledge, sometimes encoded in the myths and folklore passed down through generations." This speaks to an error in approach consistent in the reasoning of most of those who opposed the position I took in the minimization of mythos and folklore. This error takes the mythos and folklore and then views them through a literalist perspective, effectively eliminating any figurative understanding. Many stances denied the reality of usefulness in the utilization of examining mythic and folkloric stories to decode what exactly these people thought, which is relevant to our research questions.

The University of Nigeria paper goes on to state, "Yet over the centuries, this connection to the skies has been lost. Unlike some other parts of the world, many people in west Africa seem disconnected from modern astronomy, feeling they have no stake or interest in a field that seems so remote from their daily lives. At the same time, large bodies of indigenous astronomical knowledge have faded into the night, unwritten and undocumented." (Urama, 2021) This is vital to a later alternate explanation as to how it's possible to document a concept existence in one chronological timeframe within a group of people and that concept or knowledge nonexistence in later timeframes. This possibility is more logical than the black & white, either-or reasoning that happens to be fallacious. Specifically, this reasoning could be categorized as a false dilemma fallacy." Sometimes called the "either-or" fallacy, a false dilemma is a logical fallacy that presents only two options or sides when there are many options or sides. Essentially, a false dilemma presents a "black and white" kind of thinking when there are actually many shades of gray." (Excelsior Online Writing Lab,2022)

So, to imply either Griaule influenced an understanding, or the Dogon had telescopes is faulty. Just like it would be faulty to imply, either extraterrestrial beings gave them the knowledge, or the white ethnographer made it up. These are intellectually lazy explanations that oversimplify, over-exaggerate, and essentially attempt to dismiss and minimize the scholastic reality of ethno-astronomy.

If one wants to attempt to understand a people thought process, one has to look at that people's cultural worldview, and then it would be wise after that to learn or at least comb through the language. One prominent scholar whom Shawn of the MOSSI and KPRT introduced our audience to that actually made an effort to engage in that process was Thebe Medupe, who "earned a Ph.D. in physics at the University of Cape Town, is a researcher at the South African Astronomical Observatory. On top of his research on variable stars, Medupe explores cultural astronomy and historical, scientific activity in Africa. In the 2003 documentary film Cosmic Africa, Medupe visits indigenous peoples across the continent to learn about the form and significance that astronomy takes in their cultures." (Physics Today, 2006)

He informed us that he spent a few weeks with the Dogon, who demonstrated to him how relevant the stars were in their culture. He states: "One evening with the Dogons, I went with two old people to look at the stars. I asked them what was the most important constellation for them. They said the Pleiades, a star cluster, which is very important throughout the whole of Africa. The stars are used for planting and agriculture. I asked this guy [for] positions of the stars, and he gave me the rising times and positions at different times of the year. I checked with my laptop, and he was very much correct. To me, that proved he knew what he was talking about." (Physics Today, 2006) For any individual who thought the Dogon had no reason to stare at the sky or stars, this experts experience, and own words should help you do away with that thought. Planting and agriculture are vital parts of human survival across the board. Paying poor attention to the stars linked to whether you eat or not wouldn't be wise and could result in starving to death, which isn't too beneficial for any group of humans walking this planet. Another interesting note I'd like to add is how Medupe states that these people were familiar with star clusters and actually pointed one out for him.

Now, did they utilize the term Pleiades to identify it? This is unclear as no clarification is given. What is known is the Dogon have their own language, and that term we can all agree is not from any Dogon language. "Pleiades (n.) late 14c., Pliades, "visible open star cluster in the constellation Taurus," in Greek mythology, they represent the seven daughters of Atlas and Pleione, transformed by Zeus into seven stars. From Latin Pleiades, from Greek Pleiades (singular Plēias), perhaps literally "constellation of the doves" from a shortened form of peleiades, plural of peleias "dove" (from PIE root *pel- "dark-colored, gray"). Or perhaps from plein "to sail," because the season of navigation begins with their heliacal rising. Old English had the name from Latin as Pliade. The star cluster is mentioned by Hesiod (pre-700 B.C.E.); only six now are visible to most people; on a clear night, a good eye can see nine (in 1579, well before the invention of the telescope, the German astronomer Michael Moestlin (1550-1631) correctly drew 11 Pleiades stars); telescopes reveal at least 500." (Etymonline, nd)

Applying this, we now know the term derives from Greek thought, which we clearly see has it's own cultural conceptual framework. Wouldn't the Dogon have their own as well? We can deduce that Ancient Greeks didn't know it was 500 stars in that cluster, but we can see that didn't stop them from giving it a label. The etymology clearly shows a strong relationship between Greek mythology and Greek star mapping. Why is this not considered regarding the Dogon and how they mapped the stars? Why are the Greeks seemingly more superior in academia than the Dogon? The answer is that western education has a systematically and historically racist bias towards Africa and its people. It paints them as primitive savages who are too stupid to do just about anything. This sentiment can still be seen today when African Americans will get on social media platforms and argue that Africans had no concept of the universe and had no name for stars they looked at for thousands of years. I am pro-African and would never look down on my people in such a way. So, when I heard this sentiment, it made me look into the topic more especially when Sirius B entered the picture. I understood that this particular star couldn't be seen with the naked eye. That apparently didn't stop the Dogon from having the concept of Po Tolo in their mythology and including it in ritual tradition. A lot of focus and controversy is centered around the Pale fox and conversations with Ogotemmeli to assess the topic of Po Tolo. This is because this is where it's first mentioned. What I find interesting is how Griaule inserted this phrase into the Dogon vocabulary. To me, it's laughable, but some make the assertion. I personally don't know how that would work when he is being told about it by another individual. While there are many accusations, there is very little evidence to support those accusations. Though pale fox conversations with Ogotemmeli give us a reference point on the topic, I was able to read a French article translated into English called A Sudanese Sirius Star system that showed evidence of others who were familiar with this particular cluster of stars in question. The Dogon in Bandiagara, the Bambara and the Bozo in Segou, and the Minianka in Koutiala all have a "Sirius Star System" concept that was discovered.

Specifically," The main investigation was carried out among the Dogon between 1946 and 1950, where the four major informants were:

Innekouzou Dolo, a woman aged between sixty-five and seventy, ammayana 'priestess of Amma', and soothsayer, living in the Dozyou-Orey quarter of Ogol-du-Bas (Lower Ogol Sanga-du-Haut (Upper Sanga). Tribe: Arou. Language: Sanga.

Ongnonlou Dolo, between sixty and sixty-five years old, is the patriarch of the village of Go, recently established by a group of Arou in the South-West of Lower Ogol. Language: Sanga.

Yebene, fifty years old, priest of the Binou Yebene of Upper Ogol, living in Bara (Upper Sanga). Tribe: Dyon. Language: Sanga.

Manda, forty-five years old, priest of the Binou Manda, living in Orosongo in Wazouba. Tribe: Dyon. Language: Wazouba." (GRIAULE and DIETERLEN, n.d), which should clear up any misunderstanding of solely being informed by Ogotemmeli on the subject. The foreword first several paragraphs give in depth explanation of how the system was pieced together and who helped put what pieces where. We can extract from this document that "priest" played a pivotal part in obtaining knowledge on the subject. These individuals being "priests" were basically responsible for knowing pieces of the puzzle of when to celebrate a particular ritual via calculations of where certain stars were. " Every sixty years, the Dogon hold a ceremony called the Sigui (ceremony). Its purpose is the renovation of the world, and it was described at length by them in 1931." (GRIAULE and DIETERLEN, n.d) A red glow at a particular geographic location and the growth of a specific plant not sown indicate it's almost time for the ceremony.

https://www.metmuseum.org/art/collection/search/315061

The council of elders of the people of Yougo Dogorou who belong to the Arou tribe observe the signs and begin assessing "the interval by means of thirty-two-yearly drinking-bouts when beer made from millet is drunk, and the eldest elder marks up each bout with a cowrie shell."(GRIAULE and DIETERLEN, n.d)

These bouts are held in tents and shelters a month before the rain comes, sometimes in May or June. The researchers inform us that" this rule is only theoretical: between the last Sigui, celebrated at the beginning of the century, and 1931 there has been only one bout, halfway through the period; but the two-yearly cowries were set down and gathered into a pile representing the first thirty years. From 1931 onwards, the drinking bouts took place every two years. When the second pile consisting of fifteen cowries has been collected, the second Sigui of the twentieth century will be celebrated."(GRIAULE and DIETERLEN, n.d) My intent in providing these citations is to demonstrate evidence of a type of methodological formula used to calculate the ceremony and map a star's movement. This doesn't suggest an external influence, as some may have a personal interest in the topic to believe. This is also evidence of a spiritual ceremony and a type of scientific process blended together. I can agree that it may not be a bulletproof scientific formula, but I won't ignore features that involve observation and calculation but write it off as spookism. Adding to this method of calculation is the drawing of the kanaga symbol with red ochre, which represents the deity Amma, digging of a hole under the symbol, which represents the Sigui (ceremony), and the placement of a seed in the hole by elders gathered at a shelter called "tana tono" at Yougo. The researchers state, "In effect, these two signs should be 'read' In the opposite order: Amma, in the shadow of the egg (the hole) reveals himself to men (the red design) in his creative posture (the mask depicts the god's final gesture, showing the universe.)" (GRIAULE and DIETERLEN, n.d)

WE ARE THE SPEARS OF VICTORY

(Pic of mask and kanga symbol)

Interestingly the main expert who is used to refute Griaule and the knowledge of star system claim because he was unable to reproduce the same conclusion in his experiment did, in fact, encounter the kanaga symbol via mask and had this to say about in his research article titled Matter in Motion " The ritual entity, the one that moves and acts, is the material object plus the costumed dancer, and none of these elements can stand on its own. It is the combination of man, object, and act that is the real ritual agent; so, the material side of religion is another fusion of worlds, those of matter and man, inanimate and animate. Materiality is a precondition for an embodied performance, and as such can bridge the divide between the world in which we live and the 'other side,' the final border crossing."(Van Beek, 2018) He makes note of the multiple interpretations of the symbol but we can verify that both Griaule and Van Beek were able to find different Dogon tribes in Different areas with the same recurring symbol present.

A fermentation stand is also mentioned as significant in the calculation process if one is interested in more details on what was extracted.

A Sudanese Sirius System by M. GRIAULE and G. DIETERLEN is a good reference point I used here to draw from. The stands are associated with a beer-drinking ritual and are found with each regional Hogon or spiritual leader, also known as a priest. The stand woven out of baobab fibers; this stand is used during the preparation of the first ritual beer. "

This beer is distributed in small quantities to each family; it is then added to everybody's cup, thus ensuring the homogeneousness of the beer drunk by the community. In addition, all the other fermentation stands are associated, by contact, with the principal one, which is exceptionally large: the lid measures 40 cm. (16 in.) in diameter, and the four 'pompoms' are the size of the normal object. As a result, it can only enter the large jars." (Griaule & Dieterlen, n.d)

Being that all of this is associated with the Sigui it is vital to note that keeping track of the ceremony and calculation of split of the 60 yr. intervals is mnemotechnic.

"Mnemotechnics refers to the application of mnemonic principles and techniques to organize memory impressions and facilitate later recall." (Mastropieri, M.A., Scruggs, T.E., 2012). Sigi tolo or Yasig tolo (star of sigi or star of Yasig in Dogon) help demonstrate this fact by all these terms, ideas, and calculations associated with the ceremonial appearance of the star, which essentially represents the renovation of the world in Dogon ideology. This again doesn't support external influence because these aren't English terms, we translate their meaning in English because it's the language WE speak. Po means fonio (cultivated grain) in Dogon, according to a Dogon dictionary (Mombo, 2009), implying something small. Tóló, from the same dictionary, means morning star (Mombo, 2009) are apart of a system by default in the Dogon worldview and based on this ceremony. Po tóló orbits Sigi tóló is supposed to be known about by all Dogon even though the authors state that certain tribes are supposed to be held responsible for certain areas of the sky, stating, "In effect, the Ono and Domino tribes govern the stars, the former including Venus rising among its attributes, the latter Orion's belt. The sun should be assigned.

to the most powerful tribe, the Arou; but so as not to be guilty of excess, the Arou handed the sun over to the Dyon, who are less noble, and hung on to the moon." (Griaule & Dieterlen, n.d) In regards to what is stated explicitly via the Dogon on the orbit and we are informed the period of the orbit is counted double, that is, one hundred years, because the Siguis are convened in pairs of 'twins, to insist on the basic principle of twin-ness. For this reason, the trajectory is called munu, from the root monye 'to reunite,' from which the word muno is derived, which is the title given to the dignitary who has celebrated (reunited) two Siguis." (Griaule & Dieterlen, n.d)

Picture Figure iii, iv, v (Griaule & Dieterlen, n.d)

https://www.bibliotecapleyades.net/universo/siriusmystery/siriusmystery02.htm

WE ARE THE SPEARS OF VICTORY

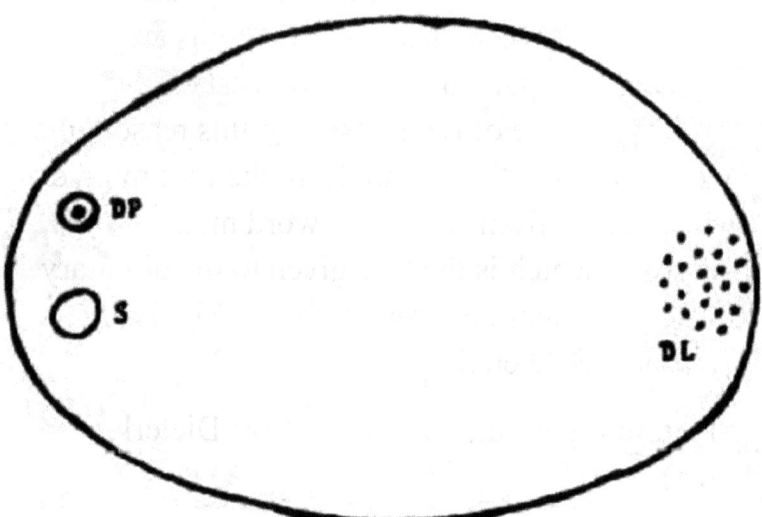

Figure iii. The trajectory of the star Digitaria around Sirius

According to Dogon mythology, the only computational Figure before the discovery of Po tóló available was that the Supreme chief was sacrificed at the end of the 7th harvest. In essence, the process is supposed to represent the regeneration of the individual "whose existence was known but whose features had not been revealed to man because the star was invisible." (Griaule & Dieterlen, n.d) So even here, the Dogon inform the curious that they aren't making a claim of seeing anything, so to say they had Supernatural vision like telescopes or that they had to have a telescope to even fathom this concept is fallacious in itself as they clearly state the star is invisible meaning they didn't say they saw anything they just had a concept attach to a visible star that describes another star orbiting it. Hence the relevance of "Po" in "Po tóló."

Here I provide an explanation of the visuals utilized to breakdown the sigui in this excerpt: "A figure made out of millet pulp (fig. iii) in the room with the dais in the house of the Hogon of Arou gives an idea of this trajectory, which is drawn horizontally: the oval (lengthwise diameter about 100 cm. = 40 in.) contains to the left a small circle, Sirius (S), above which, another circle (DP) with its center shows Digitaria in its closest position. At the other end of the oval, a small cluster of dots (DL) represents the star when it is farthest from Sirius.

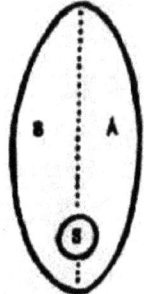

Figure iv. The symbolism of the trajectory of Digitaria. S: Sirius, A: knife, B: foreskin

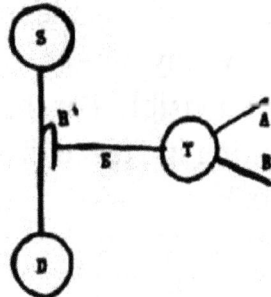

Figure v. The symbolism of Digitaria. S: Sirius, D: Digitaria, T: trajectory of Digitaria, A: knife, E: penis, B and B': foreskin

https://www.bibliotecapleyades.net/universo/siriusmystery/siriusmystery02.htm When Digitaria is close to Sirius, the latter becomes brighter; when it is at its most distant from Sirius, Digitaria gives off a twinkling effect, suggesting several stars to the observer.

This trajectory symbolizes excision and circumcision, an operation that is represented by the closest and furthest passage of Digitaria to Sirius. The left part of the oval is the foreskin (or clitoris), and the right part is the knife (fig. iv).

This symbolism is also expressed by a figure used for other performances2' (fig. v). A horizontal figure rests on a vertical axis that connects two circles: S (Sirius) and D (Digitaria); the center of the figure is a circle T, which represents the trajectory of D. Line E is the penis, the hook B' the foreskin. Two horns hinge on the circle and reproduce the two parts of the trajectory (cf. fig. iv): A, the knife; B, the foreskin." (Griaule & Dieterlen, n.d)

According to Dogon mythology, the 8th chief/Hogon who discovered the star and didn't want to get sacrificed, so he faked his death somehow and was able to lay low long enough to reveal himself to the chief who came after him so he could inform him that he had been to

Po tóló and knew all its secrets. He laid down every Hogon would reign for 60 years spill and took his place back as chief, raised the level of the sky because it had apparently been low enough to touch, and started to hate the method of calculation of time and method of reckoning. This is interesting because I would also be mad at how they calculate time because it's confusing. To demonstrate the confusing metric, this excerpt should suffice: "Until that time, the ceremonies celebrating the renovation of the world had taken place every seventh harvest; the Hogon made his calculations based on five-day periods, a unit which established the week as it still is today, and five harvest cycles. And as he was eighth in line, he counted eight cycles, in other words, forty years, and the number forty became the basis for computation: the month had forty days, the year forty weeks (of five days each).

But the Hogon lived sixty years, a number which was interpreted as the sum of forty (basis of calculation) and twenty (the twenty fingers and toes, symbolizing the person and thus, in the highest sense of the word, the chief). Thus, sixty became the basis for calculations, and it was first applied to establish the period of time

separating two Siguis. Although the orbit of Digitaria takes approximately fifty years and corresponds to the first seven reigns of seven years respectively, it nonetheless computes the sixty years which separate two ceremonies."
(Griaule & Dieterlen, n.d)

To get a clear understanding of the origin of this thought process, I again refer the reader to A Sudanese Sirius Star system, as it has an exhaustive treatment to the best of their ability. I think I have done a pretty good job in glossing over the main ideas as to how calculation comes about and who does it. I would like to add that the figurative 8th Hogon tells his people Sirius is red to the eye and po tóló white. Now this contradictory detail implies there is a way to see what is invisible to the eye when you don't take into account that this is the same mythological individual who touched and raised the level of the sky and faked his own death in the first place. We must take into consideration we are discussing a Dogon mythological account of who, what, when, where, why, and how in respect to their star system conceptualization. If he (the 8th Hogon in Dogon mythology) can touch the sky, we shouldn't be surprised he can tell you what color the invisible star is orbiting the visible one. According to him, he also went there and knew all its secrets, which apparently had to be one of them. Nevertheless, that is their explanation that doesn't scream external influence to me nor fabricated story made up by a foreigner sold as an internal account. It comes off as something organic and rationalized through Dogon lens, in my personal opinion, based on the available evidence I reviewed. I can verify that this source attempts to answer questions about what they thought the star was made out of and what they believed came from the star. It is very informative in that regard as well in touching on other tribes with Sirius Star system concepts.

To be objective in my presentation, I provided the criticism received by the school of Griaule thought and the less known and addressed critique of Van Beek. Both are prominent and respected ethnographers who compiled large amounts of information on the Dogon. What I found interesting is Van Beek may have disagreed with the findings of Griaule that he couldn't replicate. He didn't throw out Griaule work or expertise; he just challenged it how you should in a scholastic environment. Now, did he do everything to be able to replicate the experiment is what I would question. Does the fact that he couldn't find something someone else discovered mean it wasn't there and was made up? Griaule, daughter who is also an expert in the field, called Van Beek's conclusion a

"Surprise" (Calame-Griaule, 1991). She highlighted the difference in timeframe and amount of time being amongst the Dogon, and I personally think that was maybe one of Van Beek's biggest obstacle in replication. A major fair point she raises is the issue of methodology. How were these questions? Who exactly were they posed to? Was his expectation too high in thinking everyone associated with the Dogon would be familiar with in-depth knowledge of the stars when everyone in Western society can't even tell you which constellations are which, let alone how we came to the understanding of the stars in question.

She highlights from her view that Dogon would have met this questioning with distrust. This sentiment is essentially echoed in the Encyclopedia of African Religion on page 216 under Insiders and Outsiders "if a person asks a question deemed out of order, the priest is required to remain silent or even lie, to protect the inner secrets of the religion. "(Asante et al, 2008, p. 216) Why did Van Beek not consider this as a possibility is a question, I end up having in my own personal assessment in which the only answer I can come up with is either he wasn't aware or intentionally left it out. Calame-Griaule makes note of the excess amount of misreading of information given to him to be so great in number that her article doesn't have the space. This characteristic, in my view, is of someone not as closely connected to the people they are looking into than they claim. Shawn and I raised the question, was Van Beek even in the same area as Griaule? The answer is no, just like he didn't stay among the people as long as Marcel Griaule, which begs the question of how then did he expect to replicate the experiment and get an honest conclusion. Did he think he was going to do some microwave scholarship (compared to the 15+ years of Griaule) and end up with 20-year-old results? Calame-Griaule says Van Beek essentially isn't very competent in Dogon worldview, and his poor examples demonstrate this in dismissing twinness and overwhelming inaccuracies.

Another critique by a separate uninvolved scholar states, "Van Beek's were embraced by his critics and detractors, and provide a corrective to his lack of historicism and socio-political grounding, but I would argue this negative thrust is misguided." (Apter, 2005) Here the focus of Apter is the lack of understanding of a prevalent West African phenomenon of how the framework of secret societies knowledge is distributed essentially. He apparently is familiar with A Sudanese Sirius Star system as he makes mention of Griaule didn't just shoot us as knowledge seekers a paradigm anomaly because he has comparable evidence amongst the Bozo, Bambara, and Mande that I presented previously. Apter makes note of his expertise, in particular in the area of the Yoruba, and how he had to build a relationship with the people before they offered him anything information-wise. They (the Yoruba) would question his direct questions on deep knowledge, wondering why he felt he was supposed to get these knowledge-initiated individuals to have to do all kinds of stuff to obtain. This made all kinds of sense to me. Crips don't just give out their knowledge willingly neither do Free Masons. This isn't just unique to African tribes with secret societies; this sentiment exists here in our own offset of societal groupings. Why is this overlooked in these poor assessments of this topic? The most critical point I want to Highlight before I conclude is "he does not see such meaning as deep is more of a consequence of his methodical orientation: if people do not tell him it is deep, or recognize deep knowledge in an overt declaration, then in effect it does not exist for him." (Apter, 2005) regarding Van Beek. Apter is of the position that the people Van Beek asked simply either didn't know or couldn't tell him. These are reasonable and plausible assessments. Was Van Beek familiar with secret languages and how they work in these secret societies? Did he even attempt to research any terms or the language itself to see what he could find? The answers to these questions are difficult to answer based on Van Beek's research he has provided.

In Conclusion, "Stars that are grouped closely together are called star systems. Larger groups of hundreds or thousands of stars are called star clusters."
(https://courses.lumenlearning.com/geophysical/chapter/star-systems/)

Based on my assessment of linguistic evidence via secret language and lexicon suggest the Dogon had knowledge of a star system. They identify these stars Sigi Tolo, Po Tolo and emme ya. Based on the fact that the Sigi Ritual and Dogon cosmology and vocabulary predate Griaule's presence in Africa, I have not seen evidence of his influence. I have seen scholars objectively speculate, which essentially is simple conjecture that is just well articulated and well framed. What I extracted from Van Beek's work is that because he couldn't replicate, then that implies the concepts he contends did not exist. Based on the responses to Van Beek, it is clear there was a flaw in his method and reasoning, although he makes a convincing argument. I do not in any way mean to imply that Griaule's work is infallible and without flaw, but I do accept the fact that he has the most work on the topic and understands he is just the messenger ultimately relaying information given to him that he was able to document so that we can discuss such a controversial topic today. Though there may be critique in his direction, I have read nothing that completely discredits his work as a whole or the Dogon understanding.

SURVEY THREE

Syncretism: A Means and Method for Accurately Translating km.t, km.tyw, and km (Revisited)

Syncretism: A Means and Method for Accurately Translating km.t, km.tyw, and km (Revisited)

Ini Herit Shawn P

Abstract: The following is a critical analysis of the article "Syncretism: A Means and Method for Accurately Translating km.t, km.tyw, and km" by Reggie Mabry, published in the Journal of Pan-African Studies (2020). We will be examining the strengths and weaknesses of the arguments put forth by Mabry in his 2020 article. Mabry believes that the rmT (Egyptians) referred to themselves as "The Black Sacred Bull/Cow/Cattle People." After reviewing Mabry's argument and methodology, we will have uncovered several points of contention where his evidence does not support his conclusion. We will offer a few suggestions that would help him reanalyze his position to help him strengthen his overall work.

Keywords: Syncretism (Internal Syncretism), Appeal to Authority, km.t, km.tyw, km, rmT (Egyptians), Toponym, Etymology

Hypothesis: While reviewing Mabry works I it felt as if he was more concerned with attempting to defend the view of our esteemed Ancestor Diop than advancing the conversation forward. Diop has stated previously that we must not believe in his work but challenge his work for our conclusions to be held on a scientific plain. His conclusion, in my opinion, lacks the proper supportive data to substantiate his position, and his article seems to rely on syncretism without properly defining its use.

This could also explain why Mabry objected to including a working hypothesis regarding his argument, which would yield a faulty conclusion. Mabry is improperly using internal syncretism to define how the rmT 𓂋𓍿𓏏𓀀𓁐𓏪 saw themselves, and this would be an incorrect argument to make. Mabry began his article with four research questions in which he would argue his premise. Those research questions are: "What is the meaning of Kemet (km.t)? Is the term related to black and people at all? And why is Kemet (*km.t*) referred to as simply the black land by Western/European scholars? Is the meaning of Kemet important to Pan-Africanist thinking and scholarship?" (Mabry 2020:118) Here lies the focus of our analysis regarding Mabry's article, and where he did not answer his research questions by using the proper linguistic demonstration to mount his argument. As stated in my hypothesis, Mabry posted four research questions. However, we will only focus on responding to the first two questions because the way he answered his last two questions was more emotional than scholarly and repetitive in his method on how he answers the first two questions. Throughout this article, we will address his overall focus, which starts with his first question and blends the second question: "What is the meaning of Kemet (km.t), and is the term related to black people? (Mabry 2020:118) His response to the question was to look for the oldest attestation of the word km.t appearing in the sesh (hieroglyphs). He refers to Unas PT (Pyramid Text) of the 5th dynasty to explain defining the meaning of the word km.t by using internal syncretism. Let's assess his argument a little more closely, and I'll demonstrate why this is an improper argument:

WE ARE THE SPEARS OF VICTORY

Transliteration mA aHa. t(?)Unas pn m-ab ab(wy) tp.f smA(wy) n Twt is si km sA sit km.t ms.w sit bAq.t snq. w fdt wApt. (Mabry 2020:121)

Utterance 246-252: See! This Unas stands among (you), two horns are on his head [like] two wild bulls, for you are indeed the black ram, son of a black sheep, born of a bright sheep, suckled by four sheep-mothers. (Mabry 2020:121)

Mabry does not do the work to transliterate and translate this text; he relies on additional sources for his transliteration; the source he uses domain is inaccessible. He utilized pyramid text online to provide us with a sensible translation. He then argues that: "This text can be interpreted as a passage of authority. The King Unas asserts that he is a black ram, born with two horns like wild bulls, son of a black sheep born of a bright sheep. Here black ram is "si km," and black sheep is "sit km.t." We must keep in mind this is a statement of syncretism, and in the beliefs of ancient km.t civilization, this relationship existed. This relationship is not unique to how other religious figures are introduced in the text." (Mabry 2020:121)

I am not in agreement with Mabry regarding this relationship at all, and to compare the rmT ☱🕮 (Egyptians) expectations on how they did view themselves to others is not supported in his argument; however, this is how the Kings were considered.

"The ancient Egyptians regarded their King and the office of kingship as the apex and organizing principle of their society. The King's preeminent task was to preserve the right order of society, also called mAat 𓆄. This included ensuring peace and political stability, performing all necessary religious rituals, seeing to the economic needs of his people, providing justice, and protecting the country from external and internal danger. It has sometimes been said that the ancient Egyptians believed their kings to be divine, but it was the power of kingship, which the King embodied, rather than the individual himself, that was divine. The living King was associated with the God Horus and the dead King with the God Osiris, but the ancient Egyptians understood the King was mortal. One of their most ancient rituals was the Sed festival, or jubilee, at which the mortal King reaffirmed his fitness to continue as King." (Allen 2004)

So clearly, Mabry either misunderstood the source or misrepresented how internal syncretism works among the rmT ☱🕮. A perfect example of this is when he states:

"Certainly, one can make an argument that this km and km.t has nothing to do with the origins and meanings of the people of Ancient Egypt; however, they must disqualify this long tradition of the King or ruler of this civilization, stating that he was something other than a black ram or bull in the early dynasties. In other words, it would be impossible to say that km and km.t have nothing to do with the people of Ancient Egypt in color or their animal counterparts." (Mabry 2020:122)

Internal syncretism can only be argued on behalf of how the King and Queen are viewed, not the rest of the Egyptian people. If the people were considered to be just as sacred as the King and Queen, what purpose would a ruler serve among them, especially when we can assess how the rmT honored the ruling kingship as mentioned above. The rmT saw the King as divine and an extension of the gods. "In many texts, Pharaoh is called simply 'the god' ⸗ (*netjer*), or 'the good god' ⸗ (*netjer nefer*). The Egyptians, then shared, with many primitives, with the Romans, the Japanese, and the English as late as the reign of Charles II, the belief that their ruler possessed supernatural power." (Frankfort 1978:36) Here again we see internal syncretism being associated with Kinship and not the people. Mabry would supportive evidence to substantiate a stronger argument here than he did by simply basing his argument on a tradition which he misunderstands.

Mabry asserts that: "Until syncretism and a context examination are applied, this analysis asserts that km is the Black King in his Black Animal form and that Kemet (km.t) is the sacred black people in the Black animal form. At the minimum, the King and his people are represented as black bulls and cattle. What must be ruled out is that they are not Black people. Further, in this form, they are not simply black but "sacred" black. It is also safe to infer that in this form, they may be set apart from their neighbors, not in terms of a race which is a recent social construction in the historical context, but in linage." (Mabry 2020:122-123) To support this assertion, he used the same source found in Unas PT (Pyramid Text to conclude that km 𓆎𓅓𓏲 equates to Black King and now km.t 𓆎𓅓𓏏𓊖 equals The Black people. How does the association with the king to a (black) bull then transfer to the ancient Egyptian society, but also links to skin color, but is not in reference to race, but lineage? Make it make sense Reggie.

Finally, we have the answer to Mabry's first research question, and he doubles down on that answer by expanding it to refer to the people even though people were not seen as being equal to the King. No literary text or **sesh** suggests such a claim. Yet, we are supposed to assume that Mabry's attempt here to establish a connection via internal syncretism is supportive evidence that expanded beyond the kingship to the people. This is a poor argument, primarily when Mabry doesn't support his conclusion with evidence but only with assertions. He doesn't provide any linguistic or philological evidence to show when and where the rmT referred to themselves in the same manner they saw the King. What about foreign born Egyptians did this 'blackness and sacredness' apply to them as well? The single source he uses from the 5th dynastic is a funerary text venerating King Unas' journey as he travels through the afterlife looking to become a **Wsir** (Osiris), an ancestor. Therefore, it could not be referring to the people.

We will now examine PT 244-256 (Pyramid Text Online) and compare it to the citation in Mabry (2020:121).

Utterance 244 - 256:
This is here the (hard) [Eye of Horus.
His two wings have grown as those of a hawk,
(his) two feathers (are those) of a holy hawk.
His soul has brought him (here),
his magical power has adorned him.
May you open your place in heaven amongst the stars

111

WE ARE THE SPEARS OF VICTORY

of heaven!

You are indeed the unique star, the comrade of Hu.
May you look down on Osiris, when he gives orders to the spirits!
You stand high up, far from him.
You are not of them, you shall not be of them.
See! This Unas stands among (you), two horns are on his head (like) two wild bulls, for you are indeed the black ram, son of a black sheep, born of a bright sheep, suckled by four sheep-mothers.
He comes against you, Horus with blue eyes. Beware of the Horus with red eyes, whose anger is evil, whose power one cannot withstand! His messengers go, his quick runners run, they announce to the One-who-lifts-his-arm-in-the-East
that the Unique One in you is going away, (of whom) the God(?) said:
"He will give orders to (my) fathers, the gods".
The gods are silent before you, the Ennead has put their hands before their mouths, before the one in you of whom the God said:
"He will give orders to (my) fathers, the gods".
Step to the doors of the horizon, and the doors of the Cool Region open (themselves).
You stand there, ruling over them as Geb rules over his Ennead.
They come in, they strike down evil (with magical spells),
they come out, their faces are lifted up.
They see you as Min, who rules over the Two Shrines.
He stands, who stands behind you;
Your brother stands behind you;
Your relative stands behind you;
You do not go under, you will not be annihilated,

your name remains with men,
your name comes into being with the gods.

In King Unas Pyramid Text, it's not a single reference that supports Mabry's argument regarding km.t meaning "The Black People" and we are to assume that it is ok to replace sheep with bull/cattle/cow or people without any supportive linguist work to justify that assertion. However, "This view that the blood royal differs in some essential respect from ordinary men is both normal and reasonable. Without it, one cannot account for the distinction between the hereditary monarch and a usurper or the elected head of a republic. In our parlance, the usual attitude toward royalty finds expression in circumlocutions like 'His Majesty' or 'His Royal Highness.' The attitude originates, quite simply and directly, in the sense of awe - the experience of majesty undergone in the royal presence." (Frankfort 1978:36) We can see that the rmT (Egyptians) did not see themselves as Mabry's claims of equal status of the ruler, which also explains why internal syncretism is a baseless argument that does not correctly define what km or km.t means. As Mabry continues his argument and makes his assertions: "At any rate, Kemet's (km.t) people are equivalent to Kemet's (km.t) cattle in its various hieroglyphic variations, in which, only the determinatives differ. A fair question would be, in what text did the people of km.t civilization say directly that they were cattle to make the proposed relationship of cattle and people accurate?

The first case is the Westcar Papyrus story of the Dedi the Magician. In this story, the text says the following: Dd.in Hm.f imi in.tw n.i xnr nty m xnrt wd nkn.f Dd.in Ddi n is n rt(t) ity a.w.s nb.i mk n wD.tw ir.t mnt-iry n tA awt Sps.t aHa.n in n.f smn wDa(w) DADA.f aHa.n rd(.w) pA smn r gbA Said Djedi: But not to a human being, O King, my lord! Surely, it is not permitted to do such a thing to the noble cattle." (Mabry 2020:123) The source he uses to substantiate his cattle claim doesn't claim to answer the question if km.t meant cattle either. His source is referring to Inpw 𓃣 Anubis but what I found very important here is that down below, utilizing that same source, we see wt being referred to as a place name due to the O49 nwt glyph ⊗, which we see in all place names along the Nile. Mabry is making assertions about what the word km, km.t, and km.tyw mean, and none of them are based on any linguistic evidence, only false assertions of internal syncretism. I also noticed how context is missing from Mabry's argument, especially when he is relying on translations and transliterations of other sources. To save the reader the headache, km.t is a toponym which is a place name you do not travel to a people; you always travel to a place. When we book a trip to km.t, are we booking our flights to "The Black People"? No, we are not; we are planning to visit a place. Please remember km ⌃𓀀𓏥 doesn't translate as 'The King'; no dictionary entry or Egyptian text will ever yield that translation, especially when nsw.t nsw 𓇓𓏏 is the word for that term. "The Pharaoh was not mortal but a god. This was the fundamental concept of Egyptian kingship, that Pharaoh was of divine essence, a god incarnate; and this view can be traced back as far as texts and symbols take us. It is wrong to speak of a deification of Pharaoh. His divinity was not proclaimed at a certain moment, in a manner comparable to the consecration of the dead emperor by the Roman senate. His coronation was not an apotheosis but an epiphany." (Frankfort 1978:5) This is another example of the rmT 𓂋𓅓𓏏𓀀𓏥(Egyptians) not being the equivalent to the King, so, therefore, one cannot conclude that km could equal The Black King and km.t would represent The Black People.

However, Mabry triples down using a portion of a text from the 'Book of Two Ways,' and in this text, Mabry refers to the end of the text where it's talking about the nTr (God), the hidden one and all the things that he provides. Instead of giving us context, he provides a brief extraction so that he can manipulate the text to support his bias: "The third case is King Khety's instructions to his heir Merikare says - Hn rmT(t) awt nt nTr ir.n.f pt tA n ib.sn dr.n.f snk n mw ir.n.f TAw anxsfn.sn snn.f pw pr m Haw.f wbn.f m pt n ib.sn ir.n.fn.sn smw awt rm. w snm st. Provide for men, the cattle of God, for He made heaven and earth at their desire. He suppressed the greed of the waters, He gave the breath of life to their noses, for they are likenesses of Him which issued from His flesh. He shines in the sky for the benefit of their hearts; He has made herbs, cattle, and fish to nourish them. This statement from the Westcar Papyrus humans are the noble cattle (ity a.w.s nb.i), the Unas papyrus, the King is the "Black Ram born of the Black Sheep" and King Khety puts in perspective the relationship of men and cattle calling them the cattle of God (Hn rmT(t) awt nt nT)." (Mabry 2020:123)

When you examine what the text says,

"Serve God, that He may do the like for you, with offerings for replenishing the altars and with carving; it is that which will show forth your name, and God is aware of whoever serves Him. Provide for men, the cattle of God, for He made heaven and earth at their desire. He suppressed the greed of the waters, he gave the breath of life to their noses, for they are likenesses of Him which issued from His flesh. He shines in the sky for the benefit of their hearts; He has made herbs, cattle, and fish to nourish them. He has killed His enemies and destroyed His own children because they had planned to make rebellion; He makes daylight for the benefit of their hearts, and He sails around to see them, … and when they weep, He hears.… He has made for them magic to be weapons to ward off what may happen." (Encyclopedia of Religion)

We can see with more context that the end of the Book of Two Ways is not referring to a people; it refers to an offering being made in honor of the nTr ⸗ (God) for all he's done on behalf of those he's protected. Mr. Mabry is not dealing with his research questions in the manner of adequately arguing his perspective with valuable evidence. As we have seen throughout this article, he makes one assertion after another as we comb through his text. He won't even provide dictionary entrees to argue his point, only pieces of paragraphs of an overall text that has more meaning than he asserts. Let's look at just how Egyptians wrote using metaphors, praises, and adoration for the nsw.t ⸗ (King):

"The king is 'a strong bull'; a queen-mother is called 'the cow that hath borne a bull'; the sun is 'the bull of heaven'; the sky is a huge cow. A moralizing treatise states, most unexpectedly: 'Well tended are men, the cattle of God.' It is curious that scholars view such images as purely poetical without connecting them at all with the cults of Hathor, Apis, Nevis, etc. - to us equally strange. Yet both groups of phenomena obviously derive from the same root. They show that cattle played an altogether role in the consciousness of the Egyptians. This led, on the one hand, to religious venerations, and, on the other, to the spontaneous production of cattle images and cattle similes whenever some unusual observation required figurative speech for adequate expression." (Frankfort1978:162-163)

Cattle in Ancient km.t represented wealth which led to economic prowess and status among others within the society, just as in any other cattle culture throughout history. Still, it was only sacred or of religious significance to the kingship. So, any reference toward bulls as sacred, etc. are, only referring to the nTrw (Gods) or royalty in some sweet epithet. For example, we can read text describing the Sed festival and see that cattle were brought to this ceremony as a gift to the nTr (God). When the rmT refer to leadership or ownership, we can see in the language it is usually referring to a bull, yet it never refers to the people. Still, the moment we get into funerary texts such as the Pyramid or Coffin text, we can acknowledge that the rmT (Egyptians) refer to the rebirth of a new beginning in life afterlife. Therefore, the bull is also associated with Wsir 𓏏𓇋𓀭 (Osiris); he is called the Bull of the West or the Bull of Abju (Abydos). "It is clear how these indefinite beliefs could have formed the basis of the theological structures which we find in the cults of Apis, Mnevis, and Buchis. In Egypt a series of distinct markings are required for the identification of the sacred animal, while an impression of majesty suffices for the Shilluk. However, it is also clear that in both cases the sacred animals did not embody their respective gods completely or exclusively." (Frankfort 1978:167) That last line, the sacred animals did not embody their respective gods completely or exclusively, should have resonated with Mr. Mabry. He intended to draw upon and assert internal syncretism as a source for his faulty reasoning.

Regarding km.tyw, Mabry asserts that it means black cattle; his evidence for that claim is Dr. Carruthers, who uses Flinders Petrie; however, he goes a step further and accesses the Thesaurus of Linguae Aegyptiae to support the translation provided "In context and keeping syncretism in mind the km.t(j)/ km.tyw is in front of the Black Bull – Hp or Apis. What km.tyw relates to can be contrasted to tntyw.

Tntyw means those of the sacred cattle in contrast to Hwt Hr (Hathor), so kmtyw can be seen as those of the sacred black cattle in relationship to Hp (Apis)." (Mabry 2020:126)

Nowhere in his article does he have a linguistic breakdown of the word Tntyw or km.tyw to illustrate either meaning and to assert to keep in mind syncretism when referring to km.tyw is just more of the same arguments from start to finish doesn't matter which way Mabry wants to argue this topic he can't substantiate his claims with faulty logic and claim it to be scientific as he stated in his conclusion and not provide any linguistic work or full context of references he used to argue cattle, cow, or bull. This is what you see throughout his article, assertion after assertion, and when he says the source doesn't translate a word into a meaning of his choice, he follows up the translation by inserting words in places they do not translate out to be. This was, at best, a poor attempt to demonstrate support for someone else's hypothesis without even testing Diop's hypothesis out to see if he had even done the work to conclude in the matter he did. Every explanation of the mdw nTr (Medu Netcher - sacred script) in Mabry's article is not supported by the sources he used to justify its meaning.

This was no mistake on behalf of Mabry's part; this behavior was deliberate and stemmed from bias and jealous rage to go against those who were willing to do what Diop asked, and that is to not take him up on his words. Another example of Mabry projecting his bias into this discussion is when he analyzes two different translations from Egyptologist Miriam Litcheim and Renata Landgráfová. Mabry takes issue with how both Egyptologists translate the word km.tyw within the text cited. Mabry provides us without any evidence or understanding of how he arrives at his assertion versus two separate translations he refers to: "Again, returning to kmtyw, here is another example offered to resolve the meaning of kmtyw, which is the stele of Rudi Khnum and the translations of Miriam Litcheim and… In the dissertation entitled "Topic-Focus Articulation in Biographical Inscriptions and Letters of The Middle Kingdom (Dynasties11-12)" by Renata Landgráfová. She renders this as aHa.n djn.n=s w(=j) m jwn.t m wA.t wr.t n km.tjw swD.t drf xnt.t m xr.w ar.t wr.t … jw jr.n(=j) Aw.w jm=s aHa.w aA rnp.wt {r=s} n sp jj x.t nb.t jm n aA.t n rx(=j) x.t, "Then she placed me in Dendera … foremost inside the great royal palace. I spent a long time there, a long period of years {…} There never came a (bad) thing therein, for I was a knower of things." (Landgráfová) Here she does not translate km.tyw. We turn to Miriam Lichteim's "Ancient Egyptian Autobiographies chiefly of the Middle Kingdom: a study and an anthology." She translates as" Then she placed me at Dendera in her mother's (12) great cattle farm, rich in records, a foremost enterprise, the greatest estate of Upper Egypt. I have spent a long time there, a span of many years, without there being a fault (13) of mine, for my competence was great." Miriam Litcheim does not translate kmtyw to black cattle farm, but this is what the context says." (Mabry 2020:127) The way he interjects context and suggests that this is what the word is saying without definitive evidence to support his argument is proof positive that Mabry doesn't understand languages and how they work. For him to argue an entire article without any evidence to support his stance is a great example of a man with a bias and ego.

As you can see, Mabry demonstrates his argument very poorly and seems not to understand the importance of etymology and its meaning. As Durkin states: "Etymology is the investigation of word histories. It has traditionally been concerned especially with those word histories in which the facts are not certain and where a hypothesis must be constructed to account either for a word's origin or a stage in its history. That might be a stage in its meaning history, formal history, or the history of its spread from one language to another or from one group of speakers to another. The term is also used more broadly to describe the whole endeavor of attempting to provide a coherent account of a word's history (or pre-history)." (Durkin 2009:1-2) We must start here when having a linguistic conversation before we begin making assertions or trying to insinuate internal syncretism as a basis for any linguistic argument. Here is where Mabry fails to meet the mark and finds supporting linguistic work to substantiate his conclusion. He paid reverence to a 'young linguist' in his article, but he did not feel comfortable enough to rely on his work to cite his material where he claims the 'young linguist' agrees with him. To understand the importance of etymology in this conversation, we must acquire tools to help us analyze the history of a word if any changes to the word occurred over time and its meaning. I also know that we need to identify if the word possesses borrowings, sound changes, comparisons, and relationships with other languages. I know this much by studying Ranykemet 𓂋𓈖𓆎𓅓𓏏𓊖 (Egyptian Spoken Language).

"One of the most exciting aspects of etymology is that this sort of detailed work on individual word histories sometimes throws up interesting results which can have a much broader significance in tracing the history of a language (whether that be about phonology, morphology, etc.), especially when we can find parallels across a group of different word histories. Additionally, it is often crucial that questions of (non-linguistic) cultural and intellectual history are considered in tandem with questions or linguistic history." (Durkin 2009:2-3). Now that we have a brief understanding of what Etymology is and how it's used to help answer additional questions regarding languages, we can adequately assess where Mabry falls short in his argument regarding the meaning of the words km, km.t, *and* km.tyw and his methodological approach throughout. According to the TLA (Thesaurus Linguae Aegyptia), km has several different meanings: a pile of burning charcoal, black, pupil, black leather, complete, service, goal, completion, and complaint.

Vygus provides a few more entrees than the TLA: total up, wail, moan, be black, completely black bull, twinkling, profit, duty, space, and credit. I am attempting to use the same resources Mabry utilized to see if we can qualify his translation for km meaning The Black King or The King and *km.t* or *km.tyw* meaning The People or The Sacred Black People. The rmT 𓂋𓍿𓀀𓏥 has a word for people in rnykmt (RanyKemet), which is rmT, the same word for Egyptians. We have to account for the first time the rmT referred to themselves as the rmT (11th Dynasty) CT Coffin Text 1130 transliteration my own: sxpr.n.i nTrw m fdt.i rmT m rmyt n irit translation my own: I created the Gods from my sweat and humans from my tears of my eye. We also see the rmT 𓂋𓍿𓀀𓏥 referring to themselves in the Tomb of Seti I KV17 in what we know of today as 'The Table of Nations.'

Also, we can attest that the rmT 𓂋𓍿𓀀𓏥 did not start referring to the land as km.t 𓆎𓏏𓊖 until the middle kingdom. We know these things, so Mabry would have to deconstruct the language to correctly answer Diop's hypothesis, which he is arguing in favor of and assuming he is providing supportive evidence for.

So, what does the word km mean? Earlier I provided a few meanings of the word *km*, and none of them refer to a king, person, or people; it dealt with color, black, pupil, black leather, complete, service, goal, completion, and complain. The rmT has words for people and persons: rmT (people), tp ▢ (person), tpw ▢ (persons), a ▢ (person), Haw ▢ (person), rmTw ▢ (people), rmT ▢ (man, men, mankind, Egyptians, people), rmTt ▢ (men, mankind, people), wnnyw ▢ (residents, people), Xnw ▢ (people of the residence), Xtw ▢ (people), Xtw ▢ (bodies, people), nfrw ▢ (young people), xAdt ▢ (group of people), prt ▢ (people), Haw nb ▢ (all people), kAwy ▢ (other people), tmw ▢ (everyone, totality of people), and a list of other dictionary entrees. (Vygus 2018)

When we allow the rmT ▢ to speak, we do not have to rely on an assertion of syncretism to act like the rmT ▢ didn't know how to define themselves and or refer to themselves. It's the same for km.t and km.tyw as far as entrees: kmtyw ▢ (Inhabitants of Athribis), kmt ▢ (Egypt), kmt ▢ (the black land, Egypt), kmt ▢ (completion, final account), kmt ▢ (Egyptians), kmt ▢ (holy herd), and kmt ▢ (large (granite) jar, pot). (Vygus 2018) As we can see, based on a few dictionary entrees, km.t is defined as Egypt as well as The Black Land, so we must consider the argument of the Shemsu Heru Research Team. km.t is a toponym and, according to Encyclopedia Britannica: "Toponymy, a taxonomic study of place-names, based on etymological, historical, and geographical information. A place-name is a word or words used to indicate, denote, or identify a geographic locality such as a town, river, or mountain. Toponymy divides place-names into two broad categories: habitation names and feature names." (Britannica 2017)

One would have to concede that people do not travel to a people; they travel to a place, and we can see this in The Eloquent Peasant:

transliteration: mT wj m hAt r kmt r jnt aqw jm n Xrdw=j Sm swt xA n=j nA n jt. Translation: Behold, I am about to go down to Egypt to get/buy food for my children there. (Nederhof 2009) Here we have a person identifying km.t as a place and not a people. We must consider this as well as what I mentioned above regarding the importance of etymology: "Toponymy is concerned with the linguistic evolution (etymology) of place-names and the motive behind the naming of the place (historical and geographical aspects). Most Toponymy, however, has concentrated on the etymological study of habitation names, often neglecting the study of feature names and the motive behind the naming of the place." (Britannica 2017) Once we correctly understand that km.t is a place name, we can also know that the km.tyw would be a demonym that would represent the people of the place name we know as km.t. Here lies the problem with Mr. Mabry's argument he is unclear about how important place names are and how they help us identify a people regarding that geographical location.

Another example of how we can see that km.t is a place name is in the Tale of Sinuhe: transliteration sjA.n wj mTn jm pA wnn Hr kmt. Translation: Their leader, who had been in Egypt, recognized me. In the same tale, we also see the people referenced: transliteration mtr.n wj rmT kmt ntjw jm Hna=f. Translation: for the Egyptians who were there with him had borne witness to me. (Nederhof 2006) This tale is an excellent example of explaining a place's name and what demonyms. It must be considered when reviewing the sources of Mabry and the entirety of his argument; the contentions within the position he takes hop off the page. This also exposes the fallacious etymological attempts made by Mabry throughout his entire article. However, an etymological fallacy, according to Durkin is: Etymological fallacy is the idea that knowing about a word's origin, and particularly its original meaning, gives us the key to understanding its present-day use. Very frequently, this is combined with an assertion about how a word ought to be used today: certain uses are privileged as 'etymological' and hence 'valid,' while others are regarded as 'unetymological' and hence 'invalid' (or at least 'less valid'). This attitude certainly has a venerable history: the word etymology is itself ultimately from ancient Greek etumologia, which is formed from étumos' true' and logos' word speech,' hence denoting 'the study of true meanings of forms.' (Durkin 2009:27-28)

Therefore, we can honestly conclude a few things here; one: Mabry attempts to argue syncretism was a methodological mistake on his part, especially when he is unfamiliar with Egyptian grammar, Linguistics, Etymology, the meaning of Place Names, and how the rmT describe themselves. He began his approach with his bias, arguing immediately that Eurocentric scholars "are in the business of observation and creating narratives" (Mabry

2020:118); however, it is he who has stolen the European narrative, if that is so when arguing over what the words km, *k*m.t, and km.tyw mean. He didn't use a non-bias approach to tackle his four research questions. My goal here was to deal more specifically with his first two questions as it would tend to his last two questions, which I felt were emotional arguments mixed with more assertions demonstrated in his work. For every source Mabry attempted to use, he disregarded the translation that was in use only to assert that km, km.t, and km.tyw were translated correctly and meant something other than what the translator concluded with. We did not see him deconstruct the language and deal with the root of the word and its origin, nor did he show any word comparisons to substantiate his position. He merely missed the mark in his argument.

Two: Mabry's article is an attempt to Appeal to The Authority of our esteemed elders and ancestors and not do his work because he is intimidated by the task of having to go the extra mile. We can see the laziness in his article when he doesn't recreate the sesh or rely on other people's translations. When Mabry argued that two separate Egyptologists who pretty much had a word-for-word translation of a text didn't translate the word km.tyw correctly, we can immediately recognize his bias being displayed as he imposes his beliefs without using the proper tools to determine the actual meaning of a word. Three: the word km, km.t, and km.tyw has several different meanings depending on the graphemes and grammar, and to negate that was disingenuous, as we have witnessed repeatedly from his argument, and we must adhere to how etymology can help us uncover the history of a word and its meaning. Through dictionary entrees and specific primary text, we can see that in the 5th dynastic period in the Pyramid Text of Unas, the reference to 'The Black Ram, Son of a Black Sheep" is not translated as cow, cattle, or bull and is not referring to a people, and even throughout the rest of the article which continued making similar arguments which were mistakes. To the rmT, kingship was seen as godly but mortal, so he was revered in a way to possess certain qualities men (regular women and men) did not see in themselves. Cattle to the rmT were seen as sacred, represented status and determining wealth, but never represented the people. Lastly, Mabry should have known better than writing an article and concluding that others would need to use science to refute his non-scientific approach. Because he stated such, he has now professed to argue his point scientifically, and that would make his article pseudoscience because he did not demonstrate his argument scientifically.

Mabry made it clear that he would use a theoretical approach by arguing internal syncretism, which is not a science. Mabry should retract his initial argument and spend more time on learning the language properly and understanding how languages work instead of trying to finesse the readers' minds by attempting to appeal to their emotions and suggesting that his work is somehow unique because he assumes theirs no other peer review work on the subject. Mabry argues context is key, yet he ignored the context in several of the sources he used because he wanted the reader to focus only on what he is saying and trust and believe in his work to be mAa-xrw (true of voice) when it was isfet (chaotic) at best being that his arguments were all over the place.

Definitions

(1) Syncretism - Syncretism, originally a (negative) term for the eirenic theologies of Grotius (1583–1645) and Calixtus (1586–1656), was turned into a metaphor in the 1830s, apparently by J. H. Newman. Extended by C. W. King to the *Alexandrian Gnostics (1860s; see GNOSTICISM), its new meaning was summarized by Andrew Lang in relation to Egypt (1887): the word denotes the process whereby 'various god-names and god-natures are mingled so as to unite the creeds of different nomes (see NOMOS(1)) and provinces'. But the obscurity of the processes at work has meant that the term's real value lies in its imprecision. Two basic types are to be distinguished in the ancient world, 'internal' and 'contact.' **Internal syncretism is typical of ancient Egyptian (and Vedic) religion, as much the result of popular piety as of temple theology. Each God appears in a variety of forms and functions. Forms, names, and epithets diversify and intermingle with boundless energy. Gods, often in triads, co-exist or cohabit with one another, remaining separate at the level of a cult.**

(2) Appeal to Authority - The <u>fallacy</u> of appeal to authority makes the argument that if one credible source believes something that it must be true. Fallacy, Logical.

(3) Toponym - a taxonomic study of place-names based on etymological, historical, and geographical information. A place-name is a word or words used to indicate, <u>denote</u>, or identify a geographic locality such as a town, river, or mountain.

(4) Demonym - a word (such as Nevadan or Sooner) used to denote a person who inhabits or is native to a particular place.

SURVEY FOUR
The Battle of Adwa

WE ARE THE SPEARS OF VICTORY

The Battle of Adwa

Before we get into this chapter, let's discuss the Berlin Conference to establish the so-called European power's interest in the entire African continent. Now, what is the Berlin Conference, and what was discussed and agreed upon at the conference?

Berlin Conference

The Berlin Conference of 1884–1885 marked the climax of the European competition for African territory, a process commonly known as the Scramble for Africa. During the 1870s and early 1880s, European nations such as Great Britain, France, and Germany began looking to Africa for natural resources for their growing industrial sectors as well as a potential market for the goods these factories produced. As a result, these governments sought to safeguard their commercial interests in Africa and began sending scouts to the continent to secure treaties from indigenous peoples or their supposed representatives. Similarly, Belgium's King Leopold II, who aspired to increase his personal wealth by acquiring African territory, hired agents to lay claim to vast tracts of land in central Africa. To protect Germany's commercial interests, German Chancellor Otto von Bismarck, who was otherwise uninterested in Africa, felt compelled to stake claims to African land.

Appiah, A., & Gates, H. L. (2010). Encyclopedia of Africa. Oxford Univ. Press.

Berlin Conference

Inevitably, the scramble for territory led to conflict among European powers, particularly between the British and French in West Africa, Egypt, the Portuguese and British in East Africa, and the French and King Leopold II in central Africa. The rivalry between Great Britain and France led Bismarck to intervene, and in late 1884 he called a meeting of European powers in Berlin.

Otto von Bismarck was the prime minister of Prussia and the founder and first chancellor (1871–90) of the German Empire. Otto von Bismarck pursued pacific policies in foreign affairs and succeeded in preserving the peace in Europe.

Purpose of the Berlin Conference

In 1884 at the request of Portugal, German chancellor Otto von Bismark called together the major western powers of the world to negotiate questions and end confusion over the control of Africa. Bismark appreciated the opportunity to expand Germany's sphere of influence over Africa and desired to force Germany's rivals to struggle with one another for territory.

Muller, P., Hames, E. M., & Jan)., D. B. H. J. (H. (2002). Geography: Realms, regions, and concepts. J. Wiley & Sons.

Countries Represented at the Berlin Conference

Fourteen countries were represented by many ambassadors when the conference opened in Berlin on November 15, 1884. The countries represented at the time included **Austria-Hungary**, **Belgium**, **Denmark**, **France**, **Germany**, **Great Britain**, **Italy**, **the Netherlands**, **Portugal, Russia, Spain**, **Sweden-Norway** (unified from 1814-1905), **Turkey**, and **the United States of America**.

Of these fourteen nations, France, Germany, Great Britain, and Portugal was the major player in the conference, controlling most of colonial Africa.

Rosenberg, Matt. (2021, July 30). The Berlin Conference to Divide Africa. Retrieved from https://www.thoughtco.com/berlin-conference-1884-1885-divide-africa-1433556

During the carving up of Africa, you never hear much about the United States, but they were involved, and they were at the table at the conference in Berlin.

According to Peter O. Muller, At the Berlin Conference, the European colonial powers scrambled to gain control over the interior of the continent. The conference lasted until February 26, 1885 — a three-month period where colonial powers haggled over geometric boundaries in the continent's interior, disregarding the cultural and linguistic boundaries already established by the indigenous African population.

Muller, P., Hames, E. M., & Jan)., D. B. H. J. (H. (2002). Geography: Realms, regions, and concepts. J. Wiley & Sons.

During the conference, the give and take continued. By 1914, the so-called European powers had fully divided Africa among themselves into fifty countries and renamed most of them. Everyone at the conference got a piece of Africa, but Britain, France, Belgium, Portugal, Italy, Germany, and Spain were the major countries holding more of the African real estate.

1. Great Britain desired a Cape-to-Cairo collection of colonies and almost succeeded through their control of Egypt, Sudan (Anglo-Egyptian Sudan), Uganda, Kenya (British East Africa), South Africa, and Zambia, Zimbabwe

(Rhodesia), and Botswana. The British also controlled Nigeria and Ghana (Gold Coast).

2. France took much of western Africa, from Mauritania to Chad (French West Africa) and Gabon and the Republic of Congo (French Equatorial Africa).

3. Belgium and King Leopold II controlled the Democratic Republic of Congo (Belgian Congo).

4. Portugal took Mozambique in the east and Angola in the west.

5. Germany took Namibia (German Southwest Africa) and Tanzania (German East Africa).

6. Spain claimed the smallest territory — Equatorial Guinea (Rio Muni).

7. Italy's holdings were Somalia (Italian Somaliland) and a portion of Ethiopia.

Africa is rich in natural resources such as rubber, gold, diamonds, copper, bauxite, silver, petroleum, cocoa beans, iron, cobalt, uranium, and oil. These so-called Europeans wanted to rape Africa's prime real estate of its natural resources.

Now let's look into the book Africa's Natural Resources in a Global Context by Raf Custers & Ken Matthysen and see what they say about Africa's natural resources; they state Africa is high due to the continent's rich abundance of raw materials. Africa is estimated to contain 90% of the entire world's supply of platinum and cobalt, half of the world's gold supply, two-thirds of the world's manganese, and 35% of the world's uranium. It also accounts for nearly 75% of the world's coltan, an important mineral used in electronic devices, including cell phones. China has also been expanding its military presence into Africa and rivaling the United States in investment and military activity there.

Investment in the continent has also been a topic of discussion for the United States and China in their ongoing trade negotiations and political deliberations.

I want to establish to the reader why these European countries had a big interest in the entire continent of Africa.

One thing is clear the Berlin Conference established the legal claim by Europeans that all of Africa could be occupied by whoever could take it. It also established a process for Europeans to cooperate rather than fight with each other. This cooperation played a massive role in the division and conquest of Africa. It was a form of legal violence practiced upon the whole continent and all its people. For this reason, we see the Berlin Conference as a significant event in world history.

WE ARE THE SPEARS OF VICTORY

Ethiopia has abundant natural resources, such as land, timber, minerals, and gas. gold, copper, potash, platinum, and natural gas lie beneath the surface of the earth in this part of the world. During the Berlin Conference, it was agreed that Italy would get Somalia and Parts of Ethiopia, and the conference was to carve up Africa and rape it of its natural resources, so you see, Ethiopia's natural resources would make Italy a wealthy country.

Who is Menelik II

Emperor Menelik II was one of Ethiopia's greatest leaders, ruling as King and Emperor of Ethiopia from 1889 to 1913. He was born Sahle Miriam on August 17, 1844, in Ankober, Shewa, Ethiopia. His mother, Woizero Ejigayehu Lemma Adyamo, was a palace servant, and his father was Prince Haile-Melekot, Son of King Sahle Selassie. During an 1855 invasion by Emperor Tewodros II, Melekot was killed, and Miriam was taken prisoner and held captive for ten years in the emperor's mountain stronghold of Amba Magdela. He was raised alongside the Emperor's children and treated as a prince. Miriam escaped from Magdela in 1865 and returned home to Shewa. It is reported that Miriam was six feet tall and had a dark complexion with smallpox marks on his face and fine white teeth.

Nielsen, E. (2019, May 06). Emperor Menelik II (Sahle Miriam) (1844-1913). BlackPast.org.
https://www.blackpast.org/global-african-history/emperor-menelik-ii-sahle-miriam-1844-1913/

Ato Bezebeh, appointed governor by Emperor Tewodros II, fled the region when Miriam returned, and Miriam became Negus (King) of the area. When Emperor Tewodros II died in 1868, Miriam desired to become Emperor but had to submit to Tekle Giorgis (1868–1872) and Yohannes IV (1872–1889). Meanwhile, Miriam began to incorporate several kingdoms and states of southern Ethiopia into his reign, and by the time of Yohannes IV's death in 1889, Miriam had become the most powerful ruler in Ethiopia, both King, and Emperor.

Nielsen, E. (2019, May 06). Emperor Menelik II (Sahle Miriam) (1844-1913). BlackPast.org.
https://www.blackpast.org/global-african-history/emperor-menelik-ii-sahle-miriam-1844-1913/

Miriam assumed the name Emperor Menelik II upon his coronation on November 3, 1889. His ancestors had been rulers of Menz, the heartland of Shewa, since the 17th century, and it has been claimed that they were related to the Solomonid line of emperors who ruled Ethiopia between 1268 and 1854. The crown name Menelik II was significant, as Menelik, I was the legendary son of Solomon and Makeda, the Queen of Sheba. Menelik II expanded Ethiopia almost to its present-day borders.

Nielsen, E. (2019, May 06). Emperor Menelik II (Sahle Miriam) (1844-1913). BlackPast.org. https://www.blackpast.org/global-african-history/emperor-menelik-ii-sahle-miriam-1844-1913/

Menelik II said he related to King Solomon in the bible, and I do not believe King Solomon is a real historical person. Still, I do understand why he looked at or believed his lineage goes back to this Christian figure. Christianity was brought into Ethiopia via European merchants. The story of King Ezana's converting to Christianity has been reconstructed from several existing documents, the Ecclesiastical histories of Rufinus and Socrates Scholasticus. Both recount how Frementius sought Christianity from Roman merchants. King Ezana's decision to adopt Christianity was influenced by his desire to solidify his trading relationship with Roman Empire. You see this in a few cases in Africa where Kings convert to build a trading relationship with other European countries. In Africa, Kingship and chiefship often serve as spiritual and community leaders. Kings in Queens serve as high priests and high priestesses, so most of the time, when the king covert, the people covert as well.

Battle of Adwa

Decades before the Battle of Adwa, European powers had decided the fate of Ethiopia. At the Berlin Conference of 1884-1885, 14 European countries divided Africa among themselves. Before the conference, only about 10% of Africa was controlled by Europeans; the remaining 90% was ruled by indigenous and traditional rulers. Italy had colonial possession over Assab port since 1882. At the Berlin Conference, European colonial powers agreed that Italy could take over Ethiopia as its future colony.

Italy expanded its presence in the Red Sea, an area that had become important since the opening of the Suez Canal in 1869. With British support, Italy took control of the port city of Massawa in 1885. From Massawa, Italy moved slowly inland, leading to several clashes with locals, culminating in the battle of Adwa.

Yirga Gelaw Woldeyes, C. U. (n.d.). 124 years ago, Ethiopian men and women defeated the Italian Army in the Battle of Adwa. Quartz. Retrieved October 7, 2022, from https://qz.com/africa/1811232/how-ethiopians-defeated-the-italian-army-in-the-battle-of-adwa/

Italy's expansion across Ethiopia was facilitated by the devastation caused by rinderpest—an infectious viral disease—that killed up to 90% of the country's livestock. Famine and disease wiped out a third of the population between 1888 and 1892. This period is regarded as Kifu Ken, evil days.

Italy took advantage of the devastation. It sought to divide and conquer Ras Mangasha of Tigray and Nigus Menelik of Shoa. The Italians eventually signed the Treaty of Wuchale with Menelik in May 1889. The treaty was written in Amharic and Italian. The treaty would later be the trigger for the battle of Adwa. Menelik was to discover that the language in the two versions of the treaty differed. The Italian version effectively made Ethiopia Italy's protectorate, in contrast to the Amharic version.

At this point, we see that Menelik was ready to resist the Italians as they tried to gain a foothold in his territory. He had already consolidated great power and could strike once he acquired weaponry. This chapter further states that when the Italians began moving southward into Ethiopian territory, Menelik distributed weapons obtained from France and Russia, assembled a national army from Ethiopia's diverse ethnic groups, and readied his troops for battle.

Ethiopian forces

• Shewa; Negus Negasti King of Kings Menelik II: 25,000 rifles / 3,000 horses / 32 guns

• Begemder; Itaghiè Taytu: 9,000 rifles / 600 horses / 4 guns

• Gojjam; Negus Tekle Haymanot: 8,000 rifles / 700 horses

• Harar; Ras Makonnen: 15,000 rifles

• Tigray; Ras Mengesha Yohannes and Ras Alula: 5,000 rifles / 6 guns

• Wollo; Ras Mikael: 6,000 rifles / 5,000 horses

• Semien; Ras Gugsa Olié: 8,000 rifles

• Lasta; Wagshum Guangul: 6,000 rifles

- In addition, there were ~20,000 spearmen and swordsmen, as well as an unknown number of armed peasants.

McLachlan, Sean (2011). Armies of the Adowa Campaign 1896. Osprey Publishing. p. 42.

Estimates for the Ethiopian forces under Menelik range from a low of 73,000 to a high of over 120,000, outnumbering the Italians by an estimated five or six times. The forces were divided among Emperor Menelik, Empress Taytu Betul, Ras Wale Betul, Ras Mengesha Atikem, Ras Mengesha Yohannes, Ras Alula Engida (Abba Nega), Ras Mikael of

Wollo, Ras Makonnen Wolde Mikael, Fitawrari Habte Giyorgis, Fitawrari[nb, Gebeyyehu, and Negus Tekle Haymanot Tessemma. In addition, the armies were followed by a similar number of camp followers who supplied the army, as had been done for centuries. Most of the army consisted of riflemen, a significant percentage of whom were in Menelik's reserve; however, there were also a considerable number of cavalry and infantry only armed with lances (those with lances were referred to as "lancer servants")

Italian forces

The Italian army consisted of four brigades, totaling 17,978 troops with fifty-six artillery pieces. However, it is likely that fewer fought in the actual battle on the Italian side: Harold Marcus notes that "several thousand" soldiers were needed in support roles and to guard the lines of communication to the rear. He estimates that the Italian force at Adwa consisted of 14,923 effective combat troops. One brigade under General Albertone was made up of Eritrean Ascari led by Italian officers. The remaining three brigades were Italian units under Brigadiers Dabormida, Ellena, and Arimondi. While these included elite Bersaglieri and Alpini units, a large proportion of the troops were inexperienced conscripts recently drafted from metropolitan regiments in Italy into newly formed "d'Africa" battalions for service in Africa. Additionally, a limited number of troops were from the Cacciatori d'Africa, units permanently serving in Africa and, in part recruited from Italian settlers.

Raffaele Ruggeri, p. 82 Le Guerre Coloniali Italiane 1885/1900, Editrice Militare Italiana 1988

On March 1, 1896, the Ethiopian army confronted the Italians at the Battle of Adwa and scored a decisive victory. The peace treaty signed later that year preserved Ethiopia's independence during the height of the scramble for Africa.

The Treaty of Addis Ababa, signed in October 1896, did away with the Treaty of Wichale and reestablished peace. The Italian claim to control and protect Ethiopia was thereafter abandoned, and the Italian colony of Eritrea, finally fix the boundaries by a treaty of peace (September 1900), it was reduced to a territory of about 200,000 square km (80,000 square miles). Menilek's victory over the Italians gave him significant credibility with the European powers, bolstered his mandate at home, and provided the Ethiopian kingdom with a period of peace in which it was able to expand and flourish, in contrast to most of the rest of the African continent at that time, which was embroiled in colonial conflicts. Various treaties concluded with Italy, France, and Great Britain in the years up to 1908 fixed the borders of Ethiopia with the neighboring territories ruled by the European powers.

In the battle of Adwa, which took place 126 years ago, traditional warriors, farmers, pastoralists, and women, defeated a well-armed Italian army in the northern town of Adwa in Ethiopia. The outcome of this battle ensured Ethiopia's independence, making it the only African country never to be colonized. Adwa turned Ethiopia into a symbol of freedom for black people globally. It also led to a change of government in Italy.

Ethiopians celebrate their victory over the Italians every year. Historians say the victory turned Ethiopia into an icon of liberty for black people worldwide while inspiring Africans and beyond to fight for independence.

Our great historical leaders of the past, like Marcus Garvey and W.E.B. Du Bois, and others, drew inspiration from the Ethiopian victory. Several African countries adopted the green, yellow and red Ethiopian flag after colonial liberation, and a universal national anthem was created for black people.

WE ARE THE SPEARS OF VICTORY

SURVEY FIVE

Product of the 44: Extraction of liberation and empowerment through Houston's Historical African American Enclave

Product of the 44: Extraction of liberation and empowerment through Houston's Historical African American Enclave By

Chavis Tp-hsb McCray

When Kofi told us that the main theme was spears of victory, I immediately thought to try and capture a victorious historical narrative involving the community I reside in. Acres Home came to mind because I wanted to uplift a community that I witness in a somewhat negative space today.

Acres Home has rich African American history, yet if you walk down West Montgomery these days, it's rather depressing when you look at its state today. Drugs and poverty riddle the area, accompanied by occasional violence those drugs and poverty influence. What was once a booming area of African American autonomy is now a shell of what it used to be, with a plethora of abandoned houses and properties mixed in with gentrification of the area by outsiders who see potential and opportunity in the historically black area.

A major motivational factor in the development of this survey is the reality that I don't think there is much written on the history of Acres Home, so I decided to make a contribution where I saw fit. Instead of complaining about the fact, I choose to take it upon myself to put the stories of my people in this area in book form myself so that this history won't be forgotten. I believe people in the hood need to know where and what they come from to visualize where they need to go and what they need to do.

So many days, I have walked and rode these Acres home's streets and been curious about what happened here? What did this so-and-so particular building use to be used for? Why is this property not being kept up? What's the area's story? Instead of leaving these questions a mystery, I take this opportunity given to me by my brothers to be the author of the story of Acres Home for our readers who may have never heard of this place or understand its foundational significance regarding Houston's Historically African American communities go. I hope to incite a pride in its history and shine a light on the struggle that developed in the area and what type of measures our ancestors took to overcome adversity and changing times. Hopefully, I can inspire not only those who are from the area but those abroad as well as I attempt to use these words to paint and document a powerful picture of contextualized triumph in Houston's Historically African American neighborhood. I plan to take this idea and revisit the topic in a future publication, adding to it and composing other short publications on neighborhoods with similar histories.

Acres Home is my father's hood. Since a kid, my daddy used to proudly make it known that he graduated from Eisenhower High and that he was from Acres Home 44. It didn't matter if we were actually in Houston or way in San Diego. He had a black Acres Home 44 sweatshirt that he would throw on with pride on any occasion. I have always been familiar with the area because when we moved to San Diego, we would always come to visit on summer or winter break, and it was one of our first stops outside my Granny's house when we arrived. My papa's house was off of T.C jester and Little York. Little York takes you right through the heart of Acres Home when it splits into Victory, and you cross over North Shepherd. There has always been a sign there welcoming you into the neighborhood.

WE ARE THE SPEARS OF VICTORY

We always took this same route to get to my papa's house. I have pictures as a young adolescent at my papa's place in his driveway on a toy rocking horse. I had played basketball at Desoto Park (where I took Kofi and Shawn when they came out here and we met physically for the first time) and Carver Park, 2nd of the most popular parks in the area since I first started playing basketball when I was 8 years old. I have walked from my papa's house to my Granny's house on Yale and Ishmeal down the street from Studewood, aka Independent Heights (another historic black neighborhood I will cover in another publication in the future) after a bad accident where my aunt died on a road trip in San Angelo on our way here. My papa came to get my brothers and me from the hospital so we could wait for my dad because my mom was injured too seriously for us to stay with her. That day was one of my most intimate memories of the neighborhood, as it was my first time putting my feet on the concrete and exploring the hood. I had never done so in the past, and it was to me a fascinating new world that intrigued me.

At that time, no sidewalks were coming down Little York, and it was a two-lane road that was old and full of potholes. I had no idea about the history I was walking through on my way to my Granny's house. I had no idea this community was developed and maintained by my people. I would have never guessed that the woman I'd marry and bear my children with would come from this same area of traditional values, integrity, and good character. If you had told me that my cousin Marlon aka "Marlee Emg Santana" Gomez, whom I used to dribble the basketball with and freestyle with in front of my Granny's house, would rise to stardom and create an Acres Home anthem called "44 Baby" featuring one of the realist artists who is also a product of this place sometimes referred to as the Acres Shakers that would forever musically immortalize him via being the legendary artist to engrain himself into the psyche of the very thought of the 44 when it came to putting on for the hood in a major way with that "Nawfside" classic.

Lil' cuh used his words to paint a mental visual insight of a young hustler who " got it out the mud and was proud that he could say he was a 44 baby". This pride in representing this historic small area inspired me to document why people like my dad, cousin, and wife, along with others, are warranted this neighborhood pride despite its current condition. My cousin would later be gunned down while I happen to be living in Acres Home. My kids were growing up here; they were the literal embodiment of 44 babies, as it is one of their favorite songs ironically.

This work is dedicated to him and everyone else in the 44 as I am right at this very moment typing this up dead in the middle of the hood. Residing here for so long, I have the same pride even though I didn't go to M.C. Williams or Carver High with my going to school in San Diego to the end of 9th grade. I was more familiar with Studewood because that's where my Great grandparent's houses were, along with my uncle Danny whom I'll touch more on in my following publication on Studewood's history. I used to and still tell people I'm from Studewood because that is what I knew more than anything when I came to Houston, as that is where the church Bella Vista is located that I wrote about in my book Religious Beliefs Revisited the tp hsb edition. Being a long-time resident of 44 in my adult life, along with a heavy family presence today with my mom, aunt, cousin, uncles, wife, grandparents, and uncle all within walking distance from each other, I give myself the right to represent for the 44 with this survey.

Picture of Me & Marlon Gomez Long Live Marlee "EMG" Santana

"The community is located in the northwest part of Houston and is accessible via seven major thoroughfares, T.C. Jester, West Little York, West Gulf Bank, Pinemont, Victory, Ella/Wheatley, and Shepherd. It is loosely bounded by the Houston City limits and West Gulf Bank to the north; Pinemont to the south; North Shepherd and Veterans Memorial to the east, and Greater Inwood." (Super Neighborhoods, n.d.)

To offer insight into the history and development, I thought it would be appropriate to cite Acre Home chamber.com: "Acres Homes, once considered the South's largest unincorporated black community, is south of Aldine and ten miles northwest of downtown Houston in Harris County. It developed around World War I when landholders began selling off home sites in plots big enough to allow small gardens and maintain chickens or farm animals. The town derived its name from the fact that land was sold by the acre and not by the lot. The first settlers came from rural areas, attracted by the community's inexpensive land, low taxes, and the absence of city-building standards. Residents dug wells and built sanitary facilities, but conditions in the settlement subsequently declined. When

Houston approved a plan to annex the area and install water and sewer lines, Acres Homes was a 12½-square-mile, heavily wooded, dispersed slum settlement without transportation or educational facilities. Though 90 percent of the residents were homeowners, most housings was substandard."(History of Acres Home, n.d.)

Though this website ends that paragraph with emphasis on most of the housing being substandard, all kinds of people are starting to buy up and develop land Visithoustontexas.com states, "Because of its proximity to downtown and the vast amount of affordable, heavily-wooded land that still exists today in Acres Homes, local painters and sculptors have been drawn to the community. The artist's enclave has also attracted the attention of several developers, including hip-hop artist Slim Thug who grew up in Acres Homes. Other big names from Acres Homes include current Houston Mayor Sylvester Turner, award-winning actress Loretta Devine, and rappers Chamillionaire and Paul Wall. " (Acres Home, n.d.) Land developer Alfred A. Wright platted the first of several subdivisions that eventually became the African American community of Acres Homes in 1910. Wright sold parcels of varying sized to residents who were attracted to the rural area.

(Acres Home Community, 2008) Super Neighborhoods website entry highlights, Residential development occurred slowly until the late 1930s when W.W. Mount and the Wright land Company began building in the area" (Super Neighborhoods, n.d). A common theme in any source you look up on Acres Home history is the agrarian lifestyle that is still present today. People in Houston hear Acres Home and automatically think of Horses being ridden down the street and everything from chickens to goats in people's backyards. I literally can look out my window. I see a horse eating up grass on the property behind me right out my bedroom window. There is an interesting background story to how this happened after the Reconstruction era that I will expand further on in the revisiting of this subject in a later solo publication that I'm planning to add to this current writing as there is so much powerful and rich history that this survey would make me probably have the most extended entry in volume 5 if I haven't already claimed that title. For now, understand I'm just going to gloss over that part to save for the solo book.

My wife's grandfather, whom I will refer to as Papa James, is living validation of this lifestyle.

He built up his house on the property he owns off Maxroy before it was recently remodeled. He is a living testimonial of growing food and doing everything the old fashion way by trial and error as he was familiar with carpentry, still having usable tools ready to create by his hand. He is 78 yrs old and can vouch for the development of the area. I plan to transcribe an interview with the honorable elder in a future publication as he told me stories of him being a jitney and validating to me directly face-to-face the racist conditions he dealt with in his younger years, reminiscing on colored and white water fountains and having to eat at the back of restaurants along with being able to give first-hand accounts of being on public transportation before metro owned it as well as the testimony of what it was like being a jitney.

Today we have Uber and Lyft, but back in the day, you had jitneys. "Jitneys were privately owned cars that carried passengers over a regular route according to flexible schedule a way a fellow with an old Ford to make a few extra dollars by taking people to town faster and cheaper than cumbersome street cars could manage. " (Dressman, 1992). They only lasted for less than ten years in the United States, but they became pivotal to everyday life for the black man in the early 1900s. Not only was it convenient, but it was also a way to obtain autonomy financially. Houston City Council gradually abolished jitneys by 1923-24. With 9 lines operating in the city and nearly 200 estimated drivers, it significantly reduced from nearly 300 in 1918. Of the 9, the San Felipe line was the one that catered to black Houstonians whom white jitneys wouldn't let ride. " This line's jitney, owned and operated by blacks, were a part of the black community, a business somewhat on the edge of respectability and only marginally profitable that provided a needed service." (Dressman, 1992) In his essay on Black Dixie Afro Texas History and Culture in Houston, Dressman discusses the boom of jitneys and its impact on blacks in Houston. The essay details how black jitneys were shut down a year before white ones. It's a real-life story of big white streetcar companies complaining about being short-stopped by unregulated jitneys. They didn't pay taxes and took an estimated "3 million worth of business away from the trolley lines annually." (Dressman, 1992, p. 117) To be fair, there were also safety concerns because of the lack of regulations.

In 1891 electric streetcars hit Houston streets. By 1900 Houston Electric company was running things on the already 5 miles of track. But these railways were littered with issues of traffic, narrow roads, and few bridges making the environment perfect for jitneys. On pg. 118 of Black Dixie, the reasons why jitney was embraced are elaborated on. " Not only was the new mode of transportation cheap and convenient, but it also offered the means to protest the inequities of segregation on the street cars. Texas's long-standing tradition of racial separation in mass transit began with the railroads." (Dressman, 1992, p. 118)

State law required separation in 1903, so this jitney concept was a way to fight against oppression and racism. This is emphasized on pg 119 in this excerpt: "Because of deteriorating conditions within the black political situation there had been a shift in emphasis from agitation and politics to economic advancement self-help and racial solidarity often coupled with a philosophy of accommodation the development of Transportation companies, therefore, functioning three ways as a means of protesting against discrimination as a fulfillment of the dream of creating substantial negro business by an appeal to racial solidarity and hopefully as a practical solution to the transportation problems faced by the masses of boycotting Negroes" (Dressman, 1992, p. 119)

This was the Negroes Uber according to Papa James, who told me he worked as a Jitney just right around the corner from my current residence. I was intrigued by the fact that I live right around the way from places where Black power was being exercised in full force by my people. Then he mentioned the bus and how that was black owned as well. "On July 1, 1959: On this day, the state of Texas granted, under the ownership of African-Americans, a bus franchise called the Acres Homes Transit Company. This was an important day because African-Americans owned and operated a bus franchise for the first time in the history of the South." (Gonzalez, 2015) The Houston article goes on to inform the reader "Residents including teachers, businessmen, and civic leaders banded together and petitioned city hall for a permit to operate a suburban bus franchise. In a historic feat, Acres Homes Transit Company was granted a charter from the state, and the city council followed with a temporary permit to the company. Four buses driven by local drivers made forty-three round trips a day from downtown Houston to Acres Homes. It ran for eight years until the company was sold." (Gonzalez, 2015) Dora Childress highlights in her paper, "The company was developed during the height of the civil rights period. It serves as an establishment of economic empowerment during the oppressive civil rights era." (Childress, 1994) The paper further emphasizes the inequities experienced by blacks, either being denied service altogether or being treated as 2nd class citizens. "Roger Ward says bus service in Acres Homes until 1958 was provided by the white-owned Yale Street Bus Lines. Black citizens were unhappy with the segregated terms of the service and decided to drive Larry Rush, the owner of Yale Street, out of the community. A group of citizens mainly composed of ministers and educators (the growing black middle-class) formed an organization to find a public-owned transit system, the Acres Homes Transit Company. According to Ward, J.C. Cole, a minister and subsequently the company's first president, was among those who organized a jitney service while Yale Street was operating in Acres Homes." (Childress, 1994) The jitney service essentially ran the owner of the Yale Street line company out of business. "December 1958 that Yale Street stopped service to Acres Homes. In early 1959 the Acres Homes Transit Company members petitioned the city for the franchise, but to their surprise, Larry Rush also applied to run a different franchise in the area. In the court hearing, Cole recalled that the judge granted him and Rush a license to run the Acres Homes route, the notion being "may the best man win" (J. C. Cole videotaped interview 2 February 1994). Cole says the bus company had a hard beginning. The company was ordered to run five buses on the downtown route every hour. This schedule would be a problem for the company because the buses they obtained were second-hand school buses purchased for $300. each. The courts, however, allowed Rush to run the same route with only one bus and a flexible schedule (Cole 1994).

While the bus company was getting underway, the jitney service was the community's only dependable form of transportation. By July 1959, the new Acres Homes Transit Company buses began servicing the community. Members of the community who remember the buses say the ride was only twenty-five cents. And for that small sum, the buses covered a route from Acres Homes to downtown daily. The buses ran along the main road of West

Montgomery and only picked up passengers in Acres Homes." (Ward 1993) This is the same route that our today bus number 44 runs. It's interesting that the bus had 43 stops and was later called the 44 when Metro bought the company after internal financial conflict and competing with jitneys.

Though, like everything affected by time itself, the company didn't last. The people of Acres Home empowered themselves and found a solution to the inconvenience of Jim crow. Utilizing self-empowerment financially, they were able to liberate themselves and work towards financial autonomy. I am of the position this was a Spear of Victory for black people collectively and that this history should warrant pride in being a "44" Baby in the midst of its current state. Hopefully, we can draw on our ancestors' accomplishments and further them by creating more achievements. We should buy into our historic communities instead of being pushed out. We must find ways to maintain and restore our productive presence in our neighborhoods instead of folding and letting foreigners invade.

SURVEY SIX

The Attack on Intellectualism During the Age of Information

The Attack on Intellectualism During the Age of Information

Ini Herit Shawn P

Abstract: This essay will review an onslaught of attacks on intellectualism in the past few years, from issues with proper research methodology and understanding of how science works. We will set out to understand why these attacks against original research are causing such a problem for people to understand. Are we truly invested in moving information forward, or are we left to rely on said experts because they have credentials in some field of expertise? There seems to be a community of people who intend to be against the advancement of scholarly works as if its correction is unwarranted. Intellectualism has faced an issue like this before. Could history be repeating itself, or is this a ploy to personally impede progress due to problems unknown to us?

Keywords: conscious, methodology, research, science, intellectual.

The 20th century ushered in the information age with a small device that switched and controlled electrical signals and power called a transistor. This device is known as the building block of 'modern electronics.' Without this device, the first computers, as we know, would have slowed the development of this era. The transistor allowed a new age of communication to evolve and improve upon the social ingenuity and advancement of how we operate today. This process happened rapidly; for others, it couldn't have occurred better. Society took advantage of this opportunity, allowing us to access information more quickly and easily. If we look at how the development of computers, which sparked the creation of the internet, could influence an attack on intellectualism, this problem existed before the mid-twentieth century.

The age of enlightenment sparked the pursuit of knowledge all over the world. Some called this era the scientific revolution. The participants of this challenge one another using methods to falsify one's hypothesis and analysis according to data obtained through this pursuit of knowledge. One could argue that it is in this era when the rest of the world depended on the reliability of the information from philosophers and scientists alike. By this time, supremacy would invoke its dominance on what was written, taught, and known to the world. Slavery and enslavement were still a thing and as the saying goes, 'History is written by the winner' illustrated a bias that would control the narrative for centuries. In the 19th century, as people grew more aware of what was being taught and learned in academic settings, especially from the perspective of Western Academia, the rest of the world began to challenge the notion of its worldview. This posed a problem for Africa and Africans throughout the diaspora as the burden of fighting for its story to be told and left up to its colonizers and enslavers. This would be the Africans and Diasporans' opportunity to tell the world their side of the story, void of opposing ideology that had been taught since the earliest chroniclers wrote about its discovery of the Black Continent.

However, this challenge was met with fierce rebuttals by Western and non-Western adversaries because diasporans began to have access to written documentation that was once hidden from them. A united front would continue their efforts through Pan-Africanism despite communication limitations and distance. A voice would be established on paper and in arenas. The early nineteenth century scholastically had to be the most united black folk could be before egos and ignorance would cripple a movement. Ideology wasn't an issue; the primary focus was on giving Africans a voice and preventing the rest of the world from writing it out of history. Some of the best authors, researchers, and scientists armed themselves to argue against white elitists who refused to accept an African onslaught on intellectualism challenging the status quo. These efforts would continue through the nineteenth century as Eurocentric scholars would fight back in some instances and control the narrative in others.

As the 19th century progressed, African Americans had the privilege of creating organizations and research groups that would inspire the next generation to pick up the fight where the last intellectuals left off. Some progress was gained as African American scholars would lecture, write a book, create audio, and video lectures, and be joined in the fight by certain Eurocentric scholars who would credit Africans for their advancements in numerous sciences and discoveries and be the home of early Homosapien sapiens. We would see this material show up in multiple publications as science stood on the side of the African and called into question the involvement of Arab and Greco-Roman European involvement in ancient km.t, but this would lead to several intellectuals scholastically brawling over the history and identity of the rmT (Egyptians). Scholars on both sides of the argument have written rebuttals to published works creating a continuous debate that is still being had today. Today, Africans are still not responsible for most works published in Africa, no matter the field of study. According to (Falola 2022:106), Africa is responsible for 0.4% of the published results on Africa, and South Africa precisely controls 0.3% of that information. He referred to a UNESCO science report from 2005, however: "Africa contributes just 2 percent of world research output, accounts for only 1.3 percent of research spending, and produces 0.1 percent of all patents." (Signé & Gurib-Fakim 2022)

This poses a severe problem for intellectuals, especially in the vetting area of research, unless one is trained or has a level of training in the field. However, most people take issue with some intellectual's level of activity as the recent alphabet argument has created contention with the auto dictate versus said Ph.D. as if experts do not have limitations. Most said experts might have expertise in a said field unrelated to the argument. This brings us back to the initial point about how we have evolved overall in the information age and as search engines like google, safari, and more have allowed laypeople to access information very quickly at their fingertips. Due to this, social media platforms like Twitter, TikTok, Instagram, and YouTube have allowed someone uneducated in a field of study to create such confusion in a matter of 140 plus characters or an eight-hour live stream. It becomes accessible to the world almost immediately depending on the level of influence or status of a person who conveys this information; instead, whether they understand it or not.

Today the intellectual attempts to represent themselves in the image of their previous ancestors who used science to help them answer research questions and correct misinformation that has been taught in schools, published in books, or snipped and copied out of context and used as social media posts. The intellectual is seen as pro-establishment and not a member of a woke folk-conscious community because they challenge the status quo. But this is not how scholarship works, and to participate in the process, you must adhere to the same rules those who came before we have and learn how science works. A great example of intellectualism being attacked is the recent discussions that have leaked into the YouTube space. There is a research question that was posed three years ago in a publication written by Asar Imhotep and the Shemsu Heru Research team; however, this question has resurfaced by the same individuals who seem to have a personal vendetta against The Shemsu Heru Research Team instead of focusing on the research question posed by them. Countless hours of YouTube shows have been dedicated to defaming the work that each member has put into training and learning the discipline of handling the work required to answer the research question versus appealing to the authority of a scientist who did not do the job and concluding based on feeling and not evidence.

Here we have a great example of an Attack on Intellectualism: A research group proposes a question to challenge a claim that has circulated in the community for quite some time. Instead of appealing to authority, tools were required to confirm or correct the claim made in their finding through linguistic analysis and a geological explanation. A reassessment of the literature and work was also extracted to show if the actual claimant did the necessary work required to conclude the way he did, and it was verified easily that the work that needed to be done hadn't. Now we have a dilemma, an argument based on feelings and not proper research methodology. Instead of confirming the work, an attempt to salvage a hypothesis sparked an onslaught of attacks on intellectualism. Most of the people who involve themselves in conversations are handicapped. Yet, they advocate for scientific literacy and assert that someone else's work is pseudoscience because they are not as invested in the topic as others.

Comprehension is causing the bulk of the attacks on intellectualism for several reasons: 1). To require the tools to deal with the research linguistics is required. 2). You must be familiar with Ranykemet (Egyptian spoken language). 3). You must be aware of geological terms and how they are used to describe things. Now inside each step required to help you deal with the research question exist subfields to build on the knowledge needed to address the question seriously. People who oppose and argue against those who have taken the necessary steps to deal with the work state that they do not have to learn all of that nor read different work written on the subject matter because an expert who they have not vetted due to handicaps previously mention proposing otherwise without ever addressing the research question. Outside of that, people have used slander, harled insults, and even attempted to disqualify one's authorship as pseudoscience or even plagiarism. They cherry-pick information to start arguments to attack one's credibility and even assemble cliques to control the narrative. In the age of information, google has become one's friend allowing people to quickly search themselves into arguments of someone with an opposing view without having any common knowledge of the subject matter.

Controversy sales, or at least in this arena, get a lot of attention and could be a gift and a curse. The outcome has been grueling, especially when people state that they do not have to read books to understand an argument. So how can we advance our people forward if we refuse to participate in the scientific process to solve research questions that allow us the freedom to confirm history that many agree is written by the winner? This is a serious question that people who oppose having no intention of answering, only creating more confusion and being consistent in attacking the intellectuals and scientific process. What is the end game when one side of the argument is prepared to deal with the research question, and the other is invested in creating chaos? Ultimately, we will not know until google allows one to update their information and said experts agree with the research question put forth by the Shemsu Heru Research Team. Until further notice, this attack on intellectualism and the scientific process will continue to be a thing.

In closing, in the age of information where we have so much access available and the time and energy to do 8-hour live streams on a daily if not weekly targeting individuals for personal matters instead of intellectual differences has been a hindrance to the progression of what our ancestors stood for when they picked up the pen, created research teams, and began to fight to restore Africa and African history. What purpose is sitting behind a computer screen for several hours regurgitating written work you have yet to read or displaying some sort of comprehension problems that prevent you from fully understanding the bases of one's argument? No purpose is served by creating circular arguments only to default to so-called experts whose works have yet to be vetted and won't be because a learning process is required.

So, chaos is the preferred method for those who personally take issue with intellectualism, and we have now come to a misuse of the internet as a tool to disagree. It's hard to press to say if people are genuinely invested in the advancement of scholarship based on recent behaviors. However, the way information is being ignored and noise is being entertained is a fundamental cause of concern in the information age, as our ignorance will prevent the advancement of scholarship by intellectuals committed to correcting misinformation. As history repeats itself, especially within the arena of what is known as a conscious community, a person's fingers and voice no longer need to acquire the necessary tools to answer research questions. Until a conservative effort by those who claim to be a part of the learning process to understand the severity and role intellectuals play in ensuring we get things right is taken seriously. Fringe social media advocates who oppose proper methodology will continue their attacks on intellectualism during the information age.

Bibliography

Survey One

Bearden, B. R. (2006). What really happened in the jack johnson-stanley ketchel fight? Boxing. Retrieved October 30, 2022, from https://www.boxing247.com/news/bearden2006.php

Flatter, R. (n.d.). ESPN. Retrieved October 27, 2022, from http://www.espn.com/sportscentury/features/00014275.html

Gustkey, E. (1990, July 8). 80 years ago, the truth hurt : Johnson's victory over Jeffries taught lesson to White America. Los Angeles Times. Retrieved October 27, 2022, from https://www.latimes.com/archives/la-xpm-1990-07-08-sp-462-story.html

Housman, P. (2021, April 2). Jack Johnson, the fight of the century, and Race in America. American University. Retrieved October 26, 2022, from https://www.american.edu/cas/news/jack-johnson-the-fight-of-the-century-and-race-in-america.cfm

Orbach, B. Y. (2020). The johnson-jeffries fight and censorship of Black Supremacy. The New York University Journal of Law & Liberty. Retrieved October 28, 2022, from https://www.law.nyu.edu/sites/default/files/ECM_PRO_066938.pdf

UPI. (1970, January 1). Race riots in dozen cities follow Johnson Fight Victory. UPI. Retrieved October 27, 2022, from https://www.upi.com/Archives/1910/07/05/Race-riots-in-dozen-cities-follow-Johnson-fight-victory/8746818371120/

Survey Two

On the Dogon Restudied, Genevieve Calame-Griaule

Current Anthropology, Vol. 32, No. 5 (Dec., 1991), pp. 575-577

Published By: The University of Chicago Press

https://www.sciencedirect.com/science/article/abs/pii/S0278416513000743

https://courses.lumenlearning.com/geophysical/chapter/star-systems/

https://www.nytimes.com/1985/11/18/us/6th-century-manuscript-adds-to-mystery-of-star.html

https://www.acu.ac.uk/the-acu-review/celestial-stories/

Matter in Motion: A Dogon Kanaga Mask

by Walter E. A. Van Beek

Department of Cultural Studies, Faculty of Humanities, Tilburg University, Warandelaan 1, 5037 AB Tilburg, The Netherlands

Religions 2018, 9(9), 264;
https://doi.org/10.3390/rel9090264

Received: 6 August 2018 / Revised: 27 August

https://www.space.com/22509-binary-stars.htmp

https://www.bibliotecapleyades.net/universo/siriusmystery/siriusmystery02.htm

Griaule's Legacy: Rethinking "la parole claire" in Dogon Studies* - UCLA History |

Andrew Apter, 2005

Mastropieri, M.A., Scruggs, T.E. (2012). Mnemotechnics and Learning. In: Seel, N.M. (eds) Encyclopedia of the Sciences of Learning. Springer, Boston, MA. https://doi.org/10.1007/978-1-4419-1428-6_447

https://owl.excelsior.edu/argument-and-critical-thinking/logical-fallacies/logical-fallacies-false-dilemma/#:~:text=Sometimes%20called%20the%20%E2%80%9Ceither%2Dor,actually%20many%20shades%20of%20gray.

Survey Three

Archive, Egyptology. "Book No.2 Middle Egyptian Dictionary 2018 (Mark Vygus)." EGYPTOLOGY ARCHIVE, Blogger, 3 Mar. 2019, https://www.egyptologyarchive.com/2018/10/book-no2-middle-egyptian-dictionary.html.

(2) Fallacy, Logical. "Appeal to Authority." LF, 2022, https://www.logicalfallacies.org/appeal-to-authority.html.

(1) Gordon, Richard. "Syncretism." Oxford Classical Dictionary, 7 Mar. 2016, https://oxfordre.com/classics/view/10.1093/acrefore/9780199381135.001.0001/acrefore-9780199381135-e-6177.

(3) Britannica, The Editors of Encyclopaedia. "toponymy". Encyclopedia Britannica, 18 Apr. 2017, https://www.britannica.com/science/toponymy. Accessed 16 May 2022.

Guthrie, Stewart E.. "anthropomorphism". Encyclopedia Britannica, 15 Apr. 2008, https://www.britannica.com/topic/anthropomorphism. Accessed 16 May 2022.

(4Demonym." Merriam-Webster.com Dictionary, Merriam-Webster, https://www.merriam-webster.com/dictionary/demonym. Accessed 19 May. 2022.

Frankfort, Henri. Kingship and the Gods. The University of Chicago Press, 1978.

Thesaurus Linguae Aegyptiae by indicating "<pRamesseum A = pBerlin P 10499, recto: The eloquent farmer (version R) > in: TLA (<Version month/year>).

Nederhof, Mark-Jan. "Eloquent Peasant." Ancient Egyptian Texts, 17 Aug. 2009, https://mjn.host.cs.st-andrews.ac.uk/egyptian/texts/.

Nederhof, Mark-Jan. "Sinuhe." St Andrews Corpus, 4 Nov. 2006, https://mjn.host.cs.st-andrews.ac.uk/egyptian/texts/corpus/pdf/.

Durkin, Philip. The Oxford Guide to Etymology. Oxford University Press, 2009.

Unesco, Unesco. African Ethnonyms and Toponyms. United Nations Educational, Scientific and Cultural Organization, 1984.

"Egyptian Religion: An Overview ."
Encyclopedia of Religion. . Encyclopedia.com.
25 Apr. 2022
<https://www.encyclopedia.com>.

Brown, Vincent. "South Wall Hieroglyphs."
The Pyramid Texts Online - Passage
Connecting the Sarcophagus Chamber with the
Antechamber - South Wall, Pyramid Text
Online, 2015,
https://www.pyramidtextsonline.com/passageso
uth.html.

Mabry, Reginald A. "6 Syncretism- A Means
and Method for Accurately Translating KM.T,
Km ..." Syncretism- A Means and Method for
Accurately Translating KM.T, Km, Km.tyw,
Journal of Pan African Studies, Aug. 2020,
https://www.jpanafrican.org/docs/vol13no1/6%
20Syncretism-
%20A%20Means%20and%20Method%20for%
20Accurately%20Translating%20Km.t,%20Km
.tyw%20and%20Km%2013.1.pdf.

Rosmorduc, Serge. (2014). JSesh
Documentation. [online] Available at:
http://jseshdoc.qenherkhopeshef.org [Accessed
12 Jun. 2014].

"Egyptian Religion: An Overview ."
Encyclopedia of Religion. . Encyclopedia.com.
25 Apr. 2022
<https://www.encyclopedia.com>.

Allen, Susan. "Kings and Queens of Egypt." In
Heilbrunn Timeline of Art History. New York:
The Metropolitan Museum of Art, 2000–.
http://www.metmuseum.org/toah/hd/kqae/hd_k
qae.htm (October 2004)

Lichtheim, Miriam. Ancient Egyptian
Literature. First ed., vol. 1 2, University of
California Press, 1975.

Willems, Harco. A Fragment of an Early Book
of Two Ways on the Coffin of Ankh from Dayr
Al-Barshā. The Journal of Egyptian
Archaeology, 2019.

Iry-Maat, Wudjau Men-Ib. A Beginner's
Introduction to Medew Netcher: The Ancient
Egyptian Hieroglyphic System. Heka
Multimedia, 2015.

Imhotep, Asar, et al. A Contribution To The
Debate On The Meaning Of The Place-Name
Km.t. Madu-NDela Press, 2019.

Survey Four

Appiah, A., & Gates, H. L. (2010). Encyclopedia of Africa. Oxford Univ. Press.

Muller, P., Hames, E. M., & Jan)., D. B. H. J. (H. (2002). Geography: Realms, regions, and concepts. J. Wiley & Sons.

Rosenberg, Matt. (2021, July 30). The Berlin Conference to Divide Africa. Retrieved from https://www.thoughtco.com/berlin-conference-1884-1885-divide-africa-1433556

Nielsen, E. (2019, May 06). Emperor Menelik II (Sahle Miriam) (1844-1913). BlackPast.org. https://www.blackpast.org/global-african-history/emperor-menelik-ii-sahle-miriam-1844-1913/

Yirga Gelaw Woldeyes, C. U. (n.d.). 124 years ago, Ethiopian men and women defeated the Italian Army in the Battle of Adwa. Quartz. Retrieved October 7, 2022, from https://qz.com/africa/1811232/how-ethiopians-defeated-the-italian-army-in-the-battle-of-adwa/

McLachlan, Sean (2011). Armies of the Adowa Campaign 1896. Osprey Publishing. p. 42.

Raffaele Ruggeri, p. 82 Le Guerre Coloniali
Italiane 1885/1900, Editrice Militare Italiana
1988

Survey Five

https://acreshomechamber.com/history-of-acres-home/

https://www.visithoustontexas.com/about-houston/neighborhoods/acres-homes/

https://www.houstontx.gov/superneighborhoods/6.html

https://www.texasalmanac.com/places/acres-homes

https://texashistoricalmarkers.weebly.com/acres-homes-community.html

https://blog.chron.com/bayoucityhistory/2015/07/a-first-for-acres-homes-transit-company/

https://drive.google.com/file/d/1RV38bU8klugHQDPbuMBogbmKBYPn_un6/view?usp=drivesdk

https://kinder.rice.edu/2018/02/02/welcome-to-acres-homes-acres-home-acreage-homes

https://www.tshaonline.org/handbook/entries/acres-homes-transit-company

https://digital.library.unt.edu/ark:/67531/metadc279071/m1/3/

Survey Six

Castells, Manuel. The Information Age: Economy, Society, and Culture. S.n., 1996.

Conrad, Sebastian. "Enlightenment in Global History: A Historiographical Critique." OUP Academic, Oxford University Press, 21 Sept. 2012, https://academic.oup.com/ahr/article/117/4/999/33183?login=false.

Falola, Toyin. Decolonizing African Studies: Knowledge Production, Agency, and Voice. University of Rochester Press, 2022.

Britannica, The Editors of Encyclopedia. "Industrial Revolution." Encyclopedia Britannica, 23 Aug. 2022, https://www.britannica.com/event/Industrial-Revolution. Accessed 18 October 2022.

Gurib-Fakim, Ameenah, and Landry Signé. "Investment in Science and Technology Is Key to an African Economic Boom." Brookings, Brookings, 8 Mar. 2022, https://www.brookings.edu/blog/africa-in-focus/2022/01/26/investment-in-science-and-technology-is-key-to-an-african-economic-boom/#:~:text=The%20picture%20is%20particularly%20bleak,0.1%20percent%20of%20all%20patents.

WE ARE THE SPEARS OF VICTORY

WE ARE THE SPEARS OF VICTORY

WE ARE THE SPEARS OF VICTORY

WE ARE THE SPEARS OF VICTORY

www.ingramcontent.com/pod-product-compliance
Lightning Source LLC
Chambersburg PA
CBHW051925160426
43198CB00012B/2049